Green Chemistry

Studies in Philosophy, History of Ideas and Modern Societies
Edited by Jan Hartman

Volume 24

Marcin Krasnodębski

Green Chemistry
A Brief Historical Critique

Bibliographic Information published by the Deutsche Nationalbibliothek
The Deutsche Nationalbibliothek lists this publication in the Deutsche Nationalbibliografie; detailed bibliographic data is available in the internet at http://dnb.d-nb.de.

Library of Congress Cataloging-in-Publication Data
A CIP catalog record for this book has been applied for at the Library of Congress.

This work contains the results of a study conducted as part of a research project funded by the National Science Centre, project number 2019/35/D/HS3/00614, entitled "Zielona chemia avant la lettre: Historyczne i epistemologiczne fundamenty praktyk na rzecz zrównoważonego rozwoju w chemii przemysłowej (1860-1980)".

Cover illustration: Courtesy of Benjamin Ben Chaim.

ISSN 2191-1878
ISBN 978-3-631-87818-7 (Print)
E-ISBN 978-3-631-88034-0 (E-PDF)
E-ISBN 978-3-631-88047-0 (EPUB)
DOI 10.3726/b19785

© Peter Lang GmbH
Internationaler Verlag der Wissenschaften
Berlin 2022
All rights reserved.

Peter Lang – Berlin · Bern · Bruxelles · New York ·
Oxford · Warszawa · Wien

All parts of this publication are protected by copyright. Any utilisation outside the strict limits of the copyright law, without the permission of the publisher, is forbidden and liable to prosecution. This applies in particular to reproductions, translations, microfilming, and storage and processing in electronic retrieval systems.
This publication has been peer reviewed.

www.peterlang.com

List of Tables

Table 1: Some ways to think about green chemistry expressed in the literature 27
Table 2: Five most cited articles co-authored by Anastas according to Scopus (1999–2020) 32
Table 3: 12 Principles of Green Chemistry (shortened version according to the American Chemical Society) 35
Table 4: Brown vs Green Chemistry (Woodhouse and Breyman, 2005 version). 39
Table 5: Principal American green chemistries in the 1990s 73
Table 6: Presidential Green Chemistry Award recipients in 1996 83
Table 7: Topics in green chemistry at the Carnegie Mellon University, by T. Collins (1992–1995) 86
Table 8: Overview of chapters published in canonical early green chemistry books 89
Table 9: 12 principles of green chemistry (1998) 96
Table 10: Division of 12 principles of green chemistry according to their function 99
Table 11: Twelve more principles of green chemistry (Winterton, 2001) 100
Table 12: 12 principles of green engineering 102
Table 13: Principles of Green Engineering proposed during the Sandestin Conference in 2003 104
Table 14: IMPROVEMENTS PRODUCTIVELY 106
Table 15: Thirteen principles of green chemistry and engineering for GREENER AFRICA 107
Table 16: 6 principles of green extractions 108
Table 17: 12 Principles of CO2 CHEMISTRY 109
Table 18: 8 principles of green and sustainable education 109
Table 19: Most cited green chemistry articles published between 1996–2000 (in bold papers published in *Green Chemistry*; citation count as of 1st of June 2021) 145
Table 20: Most cited "other documents" in green chemistry (published between 1996 and 2020; citation count as of 1st of June 2021) 147
Table 21: Most cited green chemistry articles published between 2001–2005 (in bold papers published in *Green Chemistry*; citation count as of 1st of June 2021) 149

Table 22: Most cited green chemistry articles published between 2006–2010 (in bold papers published in *Green Chemistry*; citation count as of 1st of June 2021) 156

Table 23: Most cited green chemistry articles published between 2011–2015 (in bold papers published in *Green Chemistry*; citation count as of 1st of June 2021) 159

Table 24: Most cited green chemistry articles published between 2016–2020 (in bold papers published in *Green Chemistry*; citation count as of 1st of June 2021) 163

Table 25: Core Theses of Sanfte Chemie (soft/gentle chemistry). My clarifications in square brackets based on Fischer's comments. 199

Table 26: Rules for determining best available techniques in Annex IV to Council Directive 96/61/EC 203

Table 27: Eight Rules of Christ (for Production-Integrated Environmental Protection) 205

Table 28: Grand Challenges of Sustainability 223

Table 29: Titles of the selected papers from the 2nd Green and Sustainable Chemistry conference published in *Current Opinion in Green and Sustainable Chemistry* 232

Table 30: Seven guiding principles for sustainable chemistry (Blum et al. 2019) 236

Table 31: ISC3's key characteristics of sustainable chemistry (2021) 237

Table 32: The concept of one-world chemistry 248

Table 33: 12 principles of circular chemistry (2019) 255

Table 34: Guidelines for integrating chemistry into a circular economy 257

List of Figures

Figure 1: Evolution of the popularity of terms "environmental chemistry" and "green chemistry" in books digitized by Google according to Google Ngram (smoothing=0, English, 1965–2019) 68

Figure 2: 12 principles of green chemistry. Popularity of the term according to Google Ngram (English-speaking books, smoothing: 0, years: 1996–2019) 110

Figure 3: Number of green chemistry articles published yearly (1996–2020) according to Scopus 138

Figure 4: Most productive authors writing about green chemistry in 1996–2000 (number of articles) 142

Figure 5: Five most popular venues publishing green chemistry articles (1996–2000) 143

Figure 6: Green chemistry articles by national affiliation of the authors' institution (1996–2000) 144

Figure 7: Most popular keywords in green chemistry papers (1996–2000) 144

Figure 8: Five major journals/series publishing green chemistry articles (excluding **Green Chemistry**) between 2001 and 2005 148

Figure 9: Green chemistry articles by national affiliation of the authors' institutions (2001–2005) 149

Figure 10: Most popular keywords in green chemistry papers (2001–2005) 151

Figure 11: 5 major journals/series publishing green chemistry articles (excluding **Green Chemistry**) between 2006 and 2010 154

Figure 12: Green chemistry articles by national affiliation of the authors' institution (2006–2010) 154

Figure 13: Most popular keywords in green chemistry papers (2006–2010) 155

Figure 14: Green chemistry articles by national affiliation of the authors' institution (2011–2015) 158

Figure 15: Most popular keywords in green chemistry papers (2011–2015) 159

Figure 16: Five major journals/series publishing green chemistry articles (excluding **Green Chemistry**) between 2011 and 2015 160

Figure 17: Five major journals/series publishing green chemistry articles (excluding **Green Chemistry**) between 2016 and 2020 161

Figure 18: Green chemistry articles by national affiliation of the authors' institution (2016–2020) 162
Figure 19: Most popular keywords in green chemistry papers (2016–2020) 163
Figure 20: Popularity of terms "biomass" and "renewable" in the English speaking literature between 1950 and 2019 according to Google Ngram (smoothing = 0) 176
Figure 21: Number of papers with "biomass," "renewable feedstock" and "renewable feedstocks" in the titles, abstracts and keywords of green chemistry papers according to Scopus ... 187
Figure 22: Number of articles published yearly with "biomass" in their titles, abstracts, or keywords according to Scopus (1996–2020) 188
Figure 23: The popularity of the term bioeconomy according to Google Ngram (2000–2019) 190
Figure 24: Popularity of terms: "sustainability," "sustainable development," "biomass," and "renewable" in the English-speaking books according to Google Ngram 1960–2019 (smoothing = 0) 207
Figure 25: Popularity of "green chemistry," "life cycle assessment," and "industrial ecology" according to Google Ngram (1985–2019, smoothing = 0) 210
Figure 26: The use of the terms "sustainable chemistry" and "green chemistry" according to Google Ngram (English literature, 1990–2019, smoothing = 0) 219
Figure 27: The use of the terms "chimie durable" [sustainable chemistry] and "chimie verte" [green chemistry] according to Google Ngram (French literature, 1990–2019, smoothing = 0) 219
Figure 28: The use of the terms "nachhaltige Chemie" [sustainable chemistry] and "grüne Chemie" [green chemistry] according to Google Ngram (German literature, 1990–2019, smoothing = 0) 220
Figure 29: The use of the terms "sustainable chemistry" (nachhaltige Chemie) and the combined category of "green chemistry" + "grüne Chemie" in German-speaking books according to Google Ngram (1990–2019, smoothing = 0) 221
Figure 30: Number of papers published yearly with "sustainable chemistry" in their titles, abstracts, or keywords, according to Scopus. 225

Figure 31: Popularity of terms "circular economy," "industrial ecology" and "green chemistry" according to google Ngram (2010–2019, smoothing = 0) 251

List of Abbreviations

ACS – American Chemical Society
CMU – Carnegie Mellon University
DDT – Dichlorodiphenyltrichloroethane (insecticide)
DfE – Design for the Environment
EPA – Environmental Protection Agency
GC – Green Chemistry
GCN – Green Chemistry Network
GE – Green Engineering
GRC – Gordon Research Conferences
INRA – Institut national de la recherche agronomique (National Institute of Agricultural Research)
IOCD – International Organization for Chemical Sciences in Development
ISC3 – International Sustainable Chemistry Collaborative Centre
IUPAC – International Union of Pure and Applied Chemistry
LCA – Life-cycle assessment
MIT – Massachusetts Institute of Technology
NSF – National Science Foundation
OECD – Organisation for Economic Co-operation and Development
RCS – Royal Chemical Society
UN – United Nations

Table of Contents

List of Tables 5

List of Figures 7

List of Abbreviations 11

Funding Acknowledgements 19

Introduction 21
        A few words on methodology   23
        The input of the humanities and social sciences   25

Chapter 1: Standard narrative on the history of green chemistry 31
        1. Green chemistry: the story so far   33
        2. What does the standard narrative really tell us about green chemistry?   37
            2.1. Brown vs Green Chemistry   38
            2.2. Green chemistry versus 'command and control' approach   44
        Conclusions   47

Chapter 2: The formative 1990s 49
        1. Green chemistry outside the Anglosphere in the late 1980s and early 1990s   50
        2. Anastas's green chemistry in the US before 1998   54

2.1. Preliminary remarks	54
2.2. *Benign by Design* (1994 book)	56
2.3. *Green Chemistry. Designing Chemistry for the Environment* (1996 book)	59
2.4. Laying down the foundations of green chemistry	61
3. Non-Anastas American green chemistries	63
3.1. Hancock's environmental green chemistry	63
3.2. Garrett's toxicological green chemistry	68
3.3. Concluding remarks	72
4. Sheldon's and Trost's green chemistry metrics	74
5. Clark's Green Chemistry	79
6. But what was practised as green chemistry in the 1990s?	82
6.1. Presidential Green Chemistry Challenge Award	82
6.2. First university courses in green chemistry	85
6.3. Canonical green chemistry symposia (1994, 1996)	89
6.4. Note on the American Chemical Society Symposia Series	92
Conclusions	93

Chapter 3: 12 principles of green chemistry and their proliferation 95

1. A short epistemological introduction	95
2. Completing and reformulating the 12 principles of green chemistry	99
3. Enthronement of green chemistry principles	110
3.1. Creating the legend	110
3.2. 12 principles in green chemistry education	113
3.3. Green chemistry's self-representation	115
Conclusions	117

Chapter 4: What is green chemistry? (normative approach) 119

    1. Major controversies: chlorine sunset and fracking   119

    2. Core problem: ionic liquids   125

    3. Green chemistry metrics   129

    Conclusions: Redrawing the boundaries   133

Chapter 5: What is green chemistry? (descriptive approach) ... 137

    1. Methodology   138

    2. Birth of the discipline (1996–2000)   141

    3. Explosion of interest (2001–2005)   147

    4. New publication venues (2006–2010)   151

    5. Rise of Asia (2011–2015)   157

    6. Solidifying change (2015–2020)   160

    7. Concluding remarks   164

    8. Side note on patents.   166

Chapter 6: Biomass and doubly green chemistry 173

    1. Prehistory of principle 7   173

    2. The French connection   176

    3. Biomass and renewability in the foundational texts of green chemistry (late 1990s–early 2000s)   182

    4. The growth of the place of biomass and renewability in the literature on green chemistry   185

Chapter 7: Not only green: sustainable chemistry and past environmentally-friendly chemistries 193

    1. Forgotten alternatives   193

  1.1. Solid state-chemistry with green
    ambitions: French chimie douce 195
  1.2. Politically incorrect green chemistry:
    German sanfte Chemie 197
  1.3. Alternative pollution prevention frameworks
    in Europe 201
 2. Escaping the green: sustainable chemistry 206
  2.1. Industrial ecology (and the life-cycle assessment) 208
  2.2. 1998 OECD Sustainable Chemistry workshop 214
  2.3. German connection 217
  2.4. American trajectory: sustainable chemistry is
    green chemistry 222
  2.5. Going beyond green: early years of sustainable
    chemistry (1998–2011) 226
  2.6. Defining sustainable chemistry (2011–2017) 229
  2.7. Formalising sustainable chemistry as a discipline
    on its own (2015–2021) 234
 Conclusions 240

Chapter 8: New conceptual frontiers for chemistry and environment 243

 1. New contenders to overthrow green chemistry 244
  1.1. Conservative evolution and the risk of
    politicization of green chemistry debates 244
  1.2. One-world chemistry 246
  1.3. Circular chemistry 250
  1.4. Concluding remarks on new alternative frameworks 258
 2. New ideas for green chemistry 258
  2.1. Systems thinking and green and sustainable
    chemistry 259
  2.2. Green chemistry and social justice 263
 Conclusions 267

General conclusions: green chemistry as history of scientific ideas 271

Bibliography 275

Index of Names 295

Funding Acknowledgements

The writing of this book was funded as a part of the project "Green Chemistry avant la lettre: historical and epistemic underpinnings of sustainable practices in chemistry and chemical industry" carried out as the Sonata 16 grant No. 2019/35/D/HS3/00614 of the Polish National Science Centre.

Introduction

Green chemistry is one of the most successful sustainability-related concepts of the last few decades. New introductory textbooks devoted to this pioneering field appear on a regular basis. Academic courses or even fully fledged degrees in green chemistry are designed in an increasing number of leading global universities. The funding agencies all around the world pour hundreds of millions of dollars into green chemistry research, not to mention private R&D investments of major companies trying to 'green' their practices in the eyes of the public. Green chemistry is at the heart of long-term development plans of entire nations and international organizations such as the OECD and the United Nations. The sheer number of articles tagged with the "green chemistry" keyword is overwhelming and grows exponentially every few years. The popularity of the concept in the world of science is unquestionable.

So what exactly is green chemistry? Someone not familiar with the concept may feel that green chemistry is undoubtedly something good. The colour green is often associated with nature or the environment; green research or policy projects are the ones that revolve around health, well-being, sustainability, and positive emotions. Green chemistry is therefore certainly a 'feeling-good chemistry' and an ordinary citizen may be perfectly content with the fact that this line of research is gaining so much attention recently. If someone tried, however, to go one step further and dig deeper to find a precise definition of green chemistry, the inquirer would encounter a mass of confusing literature with definitions as numerous as they are vague. Chemists agree neither on what type of 'object' green chemistry is (discipline, subdiscipline, paradigm, method?) nor what it deals with (pollution prevention, toxicity, renewable materials, some of them, all of them?), not to mention that its relationship to other disciplines or ways of thinking (sustainability, sustainable chemistry, industrial ecology, environmental chemistry, circular economy, etc.) is far from well-established.

Green chemistry is an imbroglio of theories, methods, institutions, ideas, and especially narratives, whose meaning evolve dynamically. Green chemistry in 1996 (the year of the first Green Chemistry Challenge Award attributed by the US government) is certainly different than the one in the early 2020s, not just because of the accumulation of novel empirical results, but because the frontiers of the concept shifted. What is particularly striking about this evolution is the epistemological self-awareness of green chemistry practitioners; no other 'discipline' generated so much self-reflection on its own foundations. Indeed, chemists working in the field perfectly see difficulties with its fuzzy boundaries. Every year a cohort of new theoretical articles, introductory chapters, press publications, and official manifestos attempt to recast the history and the underpinnings of the green chemistry project in order to define it in a new way. Descriptive and normative aspects are never fully dissociated in these publications. Many authors write the history of green chemistry in a given way to justify their own vision of the of the field.

This book is not about the history of a tremendous amount of innovations and discoveries that have been often labelled as green chemistry over the last quarter-century. It does not dwell into the intricacies of research on ionic liquids, novel bio-catalysts, solventless processes, or more efficient biorefineries. Such a work on the process of 'greening' chemistry should be conducted as a collaborative project between chemists, engineers, industry representatives, and historians, and would certainly span many volumes. It would require evaluating not only scientific contributions but also their implementation by the private sector, as well as their greenness claims.

The purpose of this book is different. Its goal is to study the language used by chemists and regulators. It examines the underlying assumptions behind the scientific practice rather than scientific findings themselves. It attempts to investigate the meaning of green chemistry and sister concepts and to understand to what kind of register these frameworks belong. Is it a language of science, of ethics, or of politics? Throughout this book, I unravel the complexity of the problem and present major developments of the green chemistry narrative. This book can therefore be a helpful guide to all who want to position themselves towards green chemistry, but who are lost in all the theoretical, and often contradictory, literature produced about it. While it can certainly help newcomers to the field, it is meant to be understandable by all those interested in the politics of sustainability and in the narratives surrounding the ongoing environmental crisis, whether they have any background in chemistry or not.

All of this may sound convoluted at first, but I systematically elucidate all the problems raised above. As such, one of the main arguments of the book is that green chemistry is a distinctively twenty-first-century type of phenomenon; not merely a fad or a temporary fashion, but a certain type of scientific jargon that helps to think about problems that are simultaneously scientific and social. The ultimate goal of this work is to propose a new way of deciphering the confusing mass of scholarship produced on theoretical underpinnings of green chemistry as well as to deconstruct some of the most widespread myths about this field of knowledge.

A few words on methodology

From the methodological standpoint, the book makes use of what sociologists call the "discourse analysis," an intuitive methodology for reading texts of culture insisting on the fact that no hierarchy of sources is fixed once and for all and that every study object needs an individual appreciation to identify relevant discourse-formative elements.[1] In this book, I mostly study and analyse articles, editorials, and book chapters explicitly defining (or trying to define) green chemistry and similar concepts. As previously mentioned, these publications are very numerous. A search in the Scopus database returns around 30 000 articles with the term green chemistry in their abstracts, titles, and keywords, or in the name of the journals that published them. Even if only 1% of them tried to define green chemistry in some way, the number still amounts to 300 papers. And it does not include many important books and early articles that have not been indexed in Scopus, it does not include non-English speaking publications, and it does not include definitions from various official websites. Since the sheer number of these documents is impressive, I had to identify the protagonists of the green chemistry history, as well as the key fora where the debate takes place, in order to select the texts I focus on. Not every definition is equally influential. And yet the definitions coming from less prominent and younger scholars can also be precious in the sense that they provide us hints on how the notion was being rebuilt and appropriated over the years and may indicate where it is going in the future. There is no straightforward way to introduce all these texts at the same time and I justify the choice of this or that document on an individual basis. In a certain way, this book constitutes an informed commentary to these various scholarly works.

Two objections against this approach may be raised at this point. Firstly, it may appear somewhat artificial to separate short bits and pieces defining green chemistry from the main body of different scientific texts that often advance informed arguments on much more specialized topics. However, the purpose of my study is not to paint a broad picture of every single subject that has been labelled as green chemistry and even less to engage in much more complex scientific debates. At the heart of the endeavour is the term green chemistry itself in its multiple declinations over the years.

---

1   Fran Tonkiss, "Discourse Analysis," in *Researching Society And Culture*, ed. Clive Seale (London: Sage, 2004), 245–260.

The second objection is that discourse analysis is not a proper research methodology due to its inherent subjectivity. It is true that with hundreds of articles trying to define green chemistry, some intuitive selection of the books and papers to focus on had to be done and that these choices may be deemed controversial. However, these difficulties do not invalidate the project's ambition. If they did, there would be no way to study narratives in the history of science in the twenty-first century at all, simply because of the number of possible sources. The historian's agency proves itself crucial in such a task and should not be hidden. The book explicitly argues for a certain reading of green chemistry and opens itself for further criticism. If the readers consider my interpretation erroneous, the future dialogue over its content can be revealing for understanding the complexity of the contemporary scientific jargon. Or to put it more modestly: I am happy to be proven wrong and have my findings subjected to critique, but the discussion needs to start somewhere, and this is why this book was needed.

Is there any way to learn what green chemistry consists of other than reading explicit definitions? Perhaps the most intuitive way of discovering what hides behind the term is to study through bibliometric lenses the content of journals, reviews, and book series referring to the framework. While in the early 2000s there was only one major review, *Green Chemistry*, which set the tone, twenty years later there is more than half a dozen journals devoted to green chemistry-related subjects, not to mention many other general chemistry reviews publishing articles tagged with the "green chemistry" keyword. Scientometric tools in modern databases such as Scopus or Web of Science enable an extensive analysis of long-term tendencies in such publications. However, scientometric analyses do not overcome the major definition-related problem. Do all articles tagged or described as green chemistry genuinely enter into the perimeter of the field? One could answer that yes, that the definition of green chemistry is what people publish as green chemistry, no matter what is considered the 'canonical' green chemistry in manuals and foundational texts. Leaving aside the question whether such an approach is appropriate for analysing value-laden concepts such as green chemistry, it certainly does not solve the problem of going in the opposite direction. What about the studies that do satisfy the definition of green chemistry but are not labelled as such, or use a competing terminology such as sustainable chemistry? I clarify these challenges throughout the text, insisting on numerous tensions surrounding the dominant vocabulary. In a way, this book constitutes a guide to debates on the frontiers of green chemistry.

If someone tried to position this work among the already established methodological frameworks of history of science, it stems, to an extent, from the philosophy of the French historian Jacques Roger.[2] Roger brought into the history of science the methods of the French Annales School well-known for its work on the history of mentalities: the overarching representations of the world in which people live and with which they interact. The history of green chemistry is above all the history of the mentalities of scientists interacting with their objects of study, contextualizing them, and explaining their research choices. I could simply call it a paradigm, but the use of the word 'paradigm' in a history of science book may suggest that the author means the term in a sense given to it by Thomas Kuhn, which is not the case.[3] Green chemistry is not (or at least not yet and not only) a Kuhnian paradigm. The concept of scientific mentalities is more helpful, but again, these methodological debates should not detract us from the core narrative of the book and are not needed to understand its message.

At the same time, this book can also be seen through the lenses of the recent literature on the history of various sub-disciplines of chemistry; a topic that has been recently gaining more and more attention.[4] However, above all, it is simply a history book exploring the complex knowledge panorama of the twenty-first century, which is characterized not only by transdisciplinarity, but also by constant transgressions between the scientific and the social. The choice of a more specific reading framework would narrow down the definition of green chemistry, while the entire purpose of the book is, precisely, to indicate the plurality of meanings and the internal contradictions and paradoxes in the way the term is used.

The input of the humanities and social sciences

While the problems with the green terminology in chemistry have been thoroughly explored by scientists themselves, green chemistry also attracted a certain

---

2 Jacques Roger, *Pour une histoire des sciences à part entière, Texte établi par Claude Blanckaert* (Paris: Albin Michel, 1995).
3 Thomas S. Kuhn, *The Structure of Scientific Revolutions* (Chicago: University of Chicago Press, 1962).
4 See, for example: John W. Servos, *Physical Chemistry from Ostwald to Pauling* (Princeton: Princeton University Press, 1990); Robert W. Cahn, *The Coming of Materials Science* (Londres: Pergamon, 2001); Baptiste Voillequin, *La catalyse en France (1944–2004): Dynamiques disciplinaires et régimes de production de savoir* (Paris: Editions universitaires européennes, 2010); Pierre Teissier, *Une histoire de la chimie du solide* (Paris: Herrman, 2014); Marcin Krasnodębski, "Institut du Pin et la chimie des résines en Aquitaine (1900–1970)" (PhD diss., University of Bordeaux, 2016).

interest of scholars in STS (science and technology studies or social studies of science), historians, economists, and philosophers of science. They all tried to take a step back in order to delineate the object of their studies from a broader perspective and to contribute to the general debate on what green chemistry is and should be, often highlighting these elements which chemists tended to gloss over and ignore. Interestingly, while some of these social scientists present a rather optimistic and enthusiastic view of the field, calling it occasionally outright revolutionary,[5] others are more reserved[6] and even sceptical about the concept of green chemistry.[7] A quick observation suggests that enthusiasts were more interested in the normative power of green chemistry, whereas sceptics were those who conducted more empirically-oriented studies on the concept.

One of these studies deserves an explicit mention. In 2006, an STS scholar with a background in chemistry, Jody Roberts, wrote a Ph.D. dissertation entitled "Creating Green Chemistry: Discursive Strategies of a Scientific Movement."[8] Roberts explores the origins of the concept in the US and the way it was formulated, understood, and disseminated. He relies on an extensive bibliography, interviews, as well as on his own experiences from a green chemistry summer school in 2004. Interesting, rich, but also critical in a stimulating way, Roberts's dissertation constitutes one of the most insightful works on the history and epistemology of green chemistry up to date. In my work, I agree with his dissertation's key assertions, I build upon them, and often reuse Roberts's arguments.

---

5   Edward J. Woodhouse and Steve Breyman, "Green Chemistry as Social Movement?," *Science, Technology, & Human Values* 30 (2005): 199–222; Jean-Pierre Llored, "Towards a Practical Form of Epistemology: the Case of Green Chemistry," *Studia Philosophica Estonica* 5 (2012): 36–60; Jean-Pierre Llored and Stephane Sarrade, "Connecting the philosophy of chemistry, green chemistry, and moral philosophy," *Foundations of Chemistry* 18 (2016): 125–152.

6   J. A. Linthorst, "An overview: origins and development of green chemistry," *Foundations of Chemistry* 12 (2010): 55–68.

7   Jody Roberts, "Creating Green Chemistry: Discursive Strategies of a Scientific Movement" (PhD diss., Faculty of Virginia Polytechnic Institute and State University, 2006); Estelle Garnier, "Une approche socio-économique de l'orientation des projets de recherche en chimie doublement verte" (PhD diss., University Reims Champagne-Ardenne, 2012); Martino Nieddu, Franck-Dominique Vivien, Estelle Garnier, Christophe Bliard, "Existe-t-il réellement un nouveau paradigme de la chimie verte?," *Natures Sciences Sociétés* 22 (2014): 103–113.

8   Roberts, "Creating Green Chemistry."

However, my book also differs from Roberts's dissertation in three important respects: methodology, chronology, and geographical scope. First of all, Roberts focused on green chemistry as a 'movement' from a sociological perspective. My approach is more eclectic; broader and narrower at the same time. It is narrower because I focus more on the language used by relevant stakeholders without entering into details of social and institutional networks that shaped green chemistry, but it is also broader since I consider the social movement approach to be just one of many ways to think about the field (Table 1).

Table 1: Some ways to think about green chemistry expressed in the literature

What is green chemistry?	What does it deal with?
- social movement - scientific discipline (mature or statu nascendi) - new paradigm in chemistry - synonym for sustainable chemistry - methodology or tool in service of sustainability - buzzword hiding greenwashing	- toxic waste prevention - chemical ecotoxicology - bioresources to replace fossil feedstocks - principles of sustainability in chemistry - life-cycle assessment (LCA) and similar approaches - any combination of the above

Another difference between the two works is much more straightforward. Roberts's dissertation was defended in 2006. Between 2006 and today, green chemistry evolved and changed. Most obviously, green chemistry got much bigger, with hundreds of articles being published every year. This book makes an effort to take these changes into account and tell the entire history of the phenomenon at least until 2021. In a sense, this remark applies to many other works on green chemistry written by social scientists over the last twenty years. Scholars such as Woodhouse and Breyman (2005) and Linthorst (2010) published excellent empirical studies in which they gathered facts about the history of green chemistry and offered their insightful interpretation, but they all became already somewhat dated and need to be retold in the light of subsequent developments.[9] No matter how detailed the historical narrative one can construct on rapidly developing fields such as, for example, nanotechnology, it will require a revision in ten or fifteen years. Interestingly, because of that, the historical studies

---

9   Linthorst, "An overview: origins and development of green chemistry;" Woodhouse and Breyman, "Green Chemistry as Social Movement?."

conducted in a given time become themselves historical artefacts. The articles on the history of green chemistry written in 2005, 2010, and 2015, participate themselves in the making of the narrative and tell something about the conceptual networks that shaped the perception of the field. So will this work for future historians. At the same time, I need to underline, while my work reiterates some key observations from the works mentioned above, it does not replace them. Readers interested in going beyond this book are strongly encouraged to do so by referring to these excellent articles and dissertations. In this sense, my book acts as an authoritative guide to the already existing sociological and historical literature on green chemistry.

One point, seemingly marginal but in practice essential, should be mentioned in this context. The authors writing the history of contemporary phenomena obviously rely on the ultimate information transmission technology of our times: the internet. While social scientists often frown upon the use of websites in professional publications due to their ephemeral nature, these fluid 'internet landscapes' are of prime importance for historians of the present time. Searching the term "green chemistry" yielded different results in the US in 2005, different in Japan in 2010, and different in Belgium in 2020. The importance of internet landscapes for the formation of discourses and shared subjectivities cannot be overstated. A small change in a Wikipedia definition is often more influential than a hundred scientific publications on a given topic. In a sense, historians of the contemporary have to attempt to picture the ephemeral nature of these internet landscapes in their books to preserve them for the analyses in the years to come, precisely because these landscapes are subjected to change. In this book, this problem is most visible in the scientometric part of research in which I rely on the Scopus and Google Books databases. In fact, my results are not always concordant with the results of older studies not only due to methodological changes, but also because the attribution of keywords changes over time. If in 2040 the term sustainable chemistry supersedes green chemistry (a topic I discuss in the last chapters of the book), and Scopus or another database retroactively replaces the term green with sustainable, many of the analyses I conduct here will become impossible to reproduce. And yet, preserved in this book, they will serve as a reference point for those trying to understand how these terms have been used in the past. The same applies to Google Ngram analyses that may change depending on Google's data collection and data mining policy in the future.

The final difference between this work and Roberts's is that his dissertation was focused above all on the United States or, the bottom line, on the Anglosphere in general. Considering that English is the international language of science and that many of early debates on green chemistry took place in the US, this choice is not surprising. However, the problems that green chemistry deals with were, of course, not exclusive to the English-speaking countries. Not only did individual national settings give green chemistry different spins, but also led to the development of slightly different terminologies rooted in national languages. In particular, in this book, I speak a little bit more in detail about the German and French theoretical frameworks. France and Germany have not only strong chemical industries but also possess abounding scientific literature in their respective national languages. Notably, if Roberts's dissertation is a basic historical reference for green chemistry in the US, the works by Estelle Garnier and Martino Nieddu provide key insights into the French case.[10]

To sum up, this book relies on previous studies conducted by social scientists, but also expands them conceptually, chronologically, and geographically (linguistically). Chapter 1 tells the 'standard version' of green chemistry's history as presented, above all, by the godfather of the field Paul Anastas. In this chapter, I also show how green chemistry constructed its own identity, I identify some shortcomings of the official narrative and explain how they cast a shadow on the entire green chemistry project. Chapter 2 goes back to the 1990s and explores the origins of the green chemistry terminology through the study of the field's foundational texts. I notably show that there were different competing green chemistries advocated by different stakeholders. Chapter 3 is about the famous 12 principles of green chemistry, their interpretations, their successors, as well as about their limitations. Chapter 4 discusses the problems of tracing the frontiers of the concept of green chemistry and provides some insights into the history of green metrics. Chapter 5 is a bibliometric study of what green chemistry is according to publication patterns over the years. It explores the evolution of major themes labelled as green chemistry to show that greenness is an elusive term. Chapter 6 presents a renewable turn in green chemistry that took place in the late 2000s and early 2010s, and the subsequent focus on biomass in the green chemistry community, especially in the French context. Chapter 7 discusses older alternative frameworks to green chemistry. It presents notably a few European green chemistries *avant la lettre*, as well as the concept of sustainable chemistry, with a special focus on the German-speaking scientific literature. Finally,

---

10 Garnier, "Une approche socio-économique de l'orientation des projets de recherche en chimie doublement verte;" Martino Nieddu et al, "Existe-t-il réellement un nouveau paradigme de la chimie verte?"

chapter 8 is an introduction to new contemporary concepts trying to overthrow green chemistry and it reflects on the future trends in green chemistry itself.

Chapter 1: Standard narrative on the history of green chemistry

Before dissecting the history of green chemistry and showing its inconsistencies, it is important to understand how it has been constructed and presented throughout the last decades. Is there an unambiguous way to identify the key discourse makers who shaped the way we define the term today? In all the reconstructions of its history, there is one person who is regularly placed at the heart of the entire green chemistry enterprise: Paul Anastas. Anastas (born in 1961) graduated in chemistry from Brandeis University in 1989 and rapidly joined the ranks of the American Environmental Protection Agency (EPA), one of the largest and most influential environmental research and regulatory bodies in the world. He gained international fame after the publication, in 1998, together with the industrial chemist John C. Warner, of the seminal book *Green Chemistry: Theory and Practice*, in which the two authors presented the so-called 12 principles of green chemistry.[1] Paul Anastas is the protagonist of virtually all stories on the emergence and development of green chemistry,[2] not to mention the works of Anastas himself in which he positions himself as the inventor of the term in its modern sense.[3] I show later that the question of whether Anastas is the father of green chemistry can be nuanced depending on the adopted definition, but there is no doubt about two things: firstly, he is considered to be the one according to the vast majority of historical works, and secondly, he is unquestionably the most influential figure involved in disseminating and promoting the concept of green chemistry in the world. This can be confirmed by a simple bibliometric analysis.

---

1 Paul Anastas and John C. Warner, *Green Chemistry: Theory and Practice* (Oxford: Oxford University Press, 1998).
2 Edward J. Woodhouse and Steve Breyman, "Green Chemistry as Social Movement?," *Science, Technology, & Human Values* 30 (2005): 199–222; Jody Roberts, "Creating Green Chemistry: Discursive Strategies of a Scientific Movement" (PhD diss., Faculty of Virginia Polytechnic Institute and State University, 2006); J. A. Linthorst, "An overview: origins and development of green chemistry," *Foundations of Chemistry* 12 (2010): 55–68.
3 Paul Anastas, "Origins and Early History of Green Chemistry," in *Advanced Green Chemistry. Part 1: Greener Organic Reactions and Processes*, ed. I. T Horváth and M. Malacria (London: World Scientific, 2018).

Using the Scopus database, we can search the term "green chemistry" in titles, abstracts, and keywords associated with different books and articles. The search returns thousands of publications between 1990 and 2020. I study different tendencies in green chemistry publishing and explain the methodology in detail in chapter 5, but even a quick glance at the results reveals that Paul Anastas is in the top 10 of the most productive authors writing about green chemistry (in terms of the number of articles). However, when we compare what kind of articles have been written by these authors, Paul Anastas immediately stands out. The vast majority of the scholars in question published scientific articles properly speaking, presenting experimental results or refining existing theories. Anastas's publications are on the other hand above all conceptual. They synthesize the state of the art and outline the philosophy behind the green chemistry idea. Their titles alone convey this particular ambition (Table 2).

Table 2: Five most cited articles co-authored by Anastas according to Scopus (1999–2020)

1. Paul Anastas and Nicolas Eghbali, "Green Chemistry: Principles and Practice," *Chemical Society Reviews* 39, n. 1 (2010): 301–312.
2. Paul Anastas and Mary Kirchhoff, "Origins, current status, and future challenges of green chemistry," *Accounts of Chemical Research* 25, n. 9 (2002): 686–694.
3. Paul Anastas and Julie B. Zimmerman, "Design through the 12 principles of green engineering," *Environmental Science and Technology* 37, n. 5 (2003): 94A–101A.
4. Martyn Poliakoff, J. Michael Fitzpatrick, Trevor Farren, and Paul Anastas, "Green chemistry: Science and politics of change," *Science* 297, n. 5582 (2002): 807–810.
5. István T. Horváth and Paul Anastas, "Innovations and green chemistry," *Chemical Reviews*, 107, n. 6 (2007): 2169–73.

In comparison to other important contributors to the field of green chemistry, Anastas conducted relatively little original empirical research. He has been, however, shaping the boundaries of the concept through these programmatic papers mediating the knowledge produced by the others. It is also worth noting that his and Warner's book has been, according to Scopus, cited more than 7000 times to date, many times more than any other book labelled as green chemistry in the database. The second book on the list, about ionic liquids, had only 919 citations between its publication year (2002) and 2020.[4]

Naturally, this bibliometric analysis needs to be taken with a grain of salt. A cross-referencing of Scopus with Google Scholar and other databases gives

---

4   Robin Rogers and Kenneth Seddon, *Ionic Liquids, Industrial Applications for Green Chemistry* (Oxford: Oxford University Press, 2002).

different results (notably in terms of citations), but the orders of magnitude remain the same. Anastas is the main sculptor of the narrative concerning green chemistry. This is not to neglect many other important authors, some of whom reinforced Anastas's version of green chemistry's history, while others added nuance to it or downright contradicted it. Still, a certain number of recurring elements can be identified in the dominant narrative, and even the people who rejected them felt the necessity to do so because of the narrative's pervasiveness.

So what does this standard narrative consist of exactly? For the sake of the argument, I present its most simplified version in order to identify the salient features. Then, I explain why this narrative does not hold and why the definition of green chemistry is much less straightforward than it appears at first glance. I point out in this chapter, in a somewhat anticipatory way, numerous fundamental contradictions in the field that I explore more in detail in the rest of the book. As such, the purpose of this chapter is to prepare the ground for a more thorough historical analysis.

1. Green chemistry: the story so far

Green chemistry stories usually start with the marine biologist and environmental activist Rachel Carson and her famous *Silent Spring* published in 1962. Until then, chemistry and the chemical industry (the differentiation between the two is rarely highlighted) had developed dynamically bringing prosperity to people in industrialized nations. The chemical industry remained, however, unchecked and failed to account for the environmental problems it had caused. Carson initiated a movement mindful of the risks of chemical pollution by pointing out the dangers of pesticides, notably of the infamous compound DDT, for biodiversity. Her book led to banning DDT in many countries, as well as to the emergence of environmental activism in the United States. In Europe, the triggering event is usually considered to be the thalidomide scandal in 1961, involving the use of a drug against morning sickness among pregnant women. The drug had unfortunately tragic consequences for no less than 10 000 children, 5000 in Germany alone, who were born with missing or deformed limbs. The thalidomide tragedy and Carson's *Silent Spring* allegedly led to undermining the public trust in science with chemistry having been held responsible. As Anastas and Warner explained in 1998:

> In both of the above cases, the public was well aware that the substances in question were designed by scientists ... Despite the confidence that they had placed in the scientists,

to provide innovations for society, the public began to realize that unintended and catastrophic consequences could result from the use of chemical substances.[5]

Environmental awareness grew in America throughout the 1960s, 1970s, and 1980s due to consecutive environmental disasters such as the fire on the Cuyahoga River in 1969, the pollution of soil with toxic dioxin in the town of Times Beach in 1982, as well as the famous Love Canal case from 1978 in which an entire region was declared a disaster area and no less than one thousand families had to be relocated due to the leak of chemicals that had been dumped there for decades. Outside of the US, the most prominently cited tragedies include the Seveso disaster in Italy (1976), and especially the tragic Bhopal explosion in India (1984) that consumed many thousands of lives. All these cases were thoroughly analysed in environmental scholarship and I will not explore them here any further, but simply mention that they are being regularly cited in green chemistry origin stories. Green chemistry positions itself in the same trajectory as the environmental movements these tragedies spurred on.

The rise of environmental awareness among the wider public in the US led first to the establishment and then to the extension of competencies of the Environmental Protection Agency. According to the standard narrative on green chemistry, the problem was that the EPA and the US government focused for too long on the so-called 'end-of-pipe' solutions. They were more interested in how to clean and eliminate pollution by multiplying new more and more cumbersome regulations, than in how to intrinsically prevent it from happening altogether. It was only in 1990 when US Congress passed the Pollution Prevention Act which recognized that "there are significant opportunities for industry to reduce or prevent pollution at the source through cost-effective changes in production, operation, and raw material use."[6]

So far, there are two lines of argumentation. The standard narrative claims that, on the one hand, pollution and toxicity were inherent to the chemical industry over decades but only in the 1960s did people realise the extent of their impact on the environment, and on the other hand, that the environmental end-of-pipe logic advocated by the US federal organizations proved itself inadequate to address them. The Pollution Prevention Act created, however, a prolific ground for the formation of a new way of thinking that could solve the challenge. Anastas joined the EPA exactly at this moment. In 1991, according to his own recollection, he coined the term green chemistry to describe this new

---

5   Anastas and Warner, *Green Chemistry*, 4.
6   Cited in Anastas, "Origins and Early History of Green Chemistry," 8.

philosophy of preventing chemical wastes instead of dealing with them afterwards, by optimizing or 'greening' chemical reactions to make them produce less toxic waste.[7] At this point, the core narrative of the first publications in the field usually concludes.

The standard story about later developments of green chemistry explains that this new philosophy, initially also known as the 'benign by design' chemistry, quickly gained popularity among American chemists. Seminars organized under the auspices of the American Chemical Association led to the first groundbreaking publications in the field. Over the same time, the so-called Gordon Research Conferences became a privileged arena for discussions on its future. In 1995, the first presidential award in green chemistry was established by Bill Clinton to compensate scientists and companies implementing green solutions. In 1997, the Green Chemistry Institute was created and the following year, in 1998, Anastas and Warner published their famous book enumerating the 12 principles of green chemistry. The 12 principles are presented in the dominant narrative as a game-changer; the major turning point opening new perspectives and reshaping the chemistry landscape forever (Table 3).

Table 3: 12 Principles of Green Chemistry (shortened version according to the American Chemical Society)[8]

1. [Pollution] Prevention;
2. Atom Economy;
3. Less Hazardous Chemical Syntheses;
4. Designing Safer Chemical;
5. Safer Solvents and Auxiliaries;
6. Design for Energy Efficient;
7. Use of Renewable Feedstocks;
8. Reduce Derivatives;
9. Catalysis;
10. Design for Degradation;
11. Real-time analysis for Pollution Prevention;
12. Inherently Safer Chemistry for Accident Prevention.

---

7   Anastas, "Origins and Early History of Green Chemistry," 8.
8   12 Principles of Green Chemistry on the official website of the American Chemical Society (accessed 23/05/2022) https://www.acs.org/content/acs/en/greenchemistry/principles/12-principles-of-green-chemistry.html.

The concept of green chemistry gained traction precisely after 1998. In 1999, the first journal in green chemistry, entitled simply *Green Chemistry*, was established in the UK by the British chemist James Clark. Clark remains probably the most influential leader in the field after Anastas. In the following years, seminars and conferences in green chemistry multiplied and the chapters of the Green Chemistry Institute and other green chemistry associations were established all over the world. Furthermore, throughout the 2000s, the number of articles published in the *Green Chemistry* journal grew exponentially every few years. Other journals were established as well, for example, *Green Chemistry Letters and Reviews* in 2007, *ChemSusChem* in 2008, a more polemical *Current Opinion in Green and Sustainable Chemistry* in 2016, but all of this is just the tip of the iceberg. New textbooks and book series have been published every single year and the use of the green chemistry keyword exploded over the last twenty years in the general chemistry journals. The sheer number of publications seems to confirm the success story of the green chemistry concept.

In 2016, the *Green Chemistry* journal celebrated the 25th anniversary of the concept. The 25th because, according to the editors, green chemistry was born in 1991, the year in which Anastas allegedly coined the term.[9] Throughout the entire year 2016, the journal published every month an article by a renowned scholar reflecting on one (or more) of the 12 principles and commenting on its past successes and on its importance in opening new lines of research. Two years later, in 2018, another anniversary was celebrated: the 20 years of the 12 principles.[10] In a beautifully illustrated and extensive article on the achievements of the concept, Anastas and his colleagues drew a metaphorical 'green tree model.' Every principle was a branch on the green chemistry tree. From every branch stemmed multiple smaller branches corresponding to groundbreaking innovations and promising research themes. The article explained, principle by principle, its contributions to making chemistry safer and more environmentally friendly. All these tremendous advancements were supposedly possible thanks to the 12 principles laid down by Anastas and Warner in 1998.

---

9   Paul Anastas, Buxing Han, Walter Leitner, Walter Leitner and Martyn Poliakoff, "'Happy silver anniversary:' Green Chemistry at 25," *Green Chemistry* 18 (2016): 12–13.
10 Hanno C. Erythropel et al., "The Green ChemisTREE: 20 years after taking root with 1 the 12 Principles," *Green Chemistry* 20 (2018): 1929–1961.

2. What does the standard narrative really tell us about green chemistry?

The story so far goes then this way: the chemical industry was a benefactor of humanity, but due to carelessness and the lack of understanding of potential dangers, its activities led to environmental disasters and mass pollution. Rachel Carson was one the first to react and paved the way to environmental pollution regulation. But the end-of-pipe mentality dominant in the 1970s and 1980s was insufficient and imperfect. What was necessary was the redesign of chemical processes to prevent pollution altogether. If Carson was the mother of environmentalism, Anastas was the father of this new type of environmentally friendly green chemistry codified in the 12 principles. After their formulation, the field flourished uninterrupted with hundreds of new publications every year. The ingenuity of its principles brought a whole range of discoveries and innovations making our future better and safer.

Of course, this story is crudely simplified and Anastas himself is much more nuanced, especially in his later papers, but this core narrative is what a newcomer to the field may infer from multiple texts on the history of green chemistry. In all of these texts, however, there is an underlying tension between two definitions of green chemistry that remains non-verbalized and yet casts a shadow on the success story presented above. Green chemistry is either seen as a discipline (with its own set of institutions, journals, and practices) or as a paradigm (an attribute of chemistry in general as practised in a given period). Green chemistry positions itself sometimes as one and sometimes as the other. In this book, I argue it is neither, but it certainly possesses traits of both. Jody Roberts points out in his 2006 dissertation that

> In personal communications with two core proponents of the field, I heard this same line: if we're still talking about green chemistry in 20 years then we've failed. Failed because green chemistry would still be on the outside, something else, a different way of doing things, but not the way of practicing chemistry.[11]

Elsewhere he continues:

> When all chemists think like green chemists there will be no need for a separate green chemistry. Chemistry itself will have already incorporated these principles into its practices. Yet, to transform chemistry and make it green, green chemistry itself has to become something without becoming something permanently separate, distinct, or

---

11 Roberts, "Creating Green Chemistry," 95.

other. Thus its success in creating a space for itself may result in its greatest challenge—disappearing again.[12]

More than 15 years after this statement, while it is very arguable whether green chemistry became the new paradigm of chemistry in general, one thing is sure, it failed to disappear. On the contrary, it is present more than ever. As a consequence, the success story presented in the previous section (new journals, textbooks, and thousands of articles using green chemistry as a keyword) can be seen as a failure to make green chemistry's core ambition a reality. However, if green chemistry is not a default paradigm for chemistry in general, it does not automatically mean that it is a scientific discipline properly speaking either.

How to understand this unclear epistemological status? One of the arguments of my book is that the reason for green chemistry's success in establishing institutions and spreading throughout the world of science, but at the same time for its inability to revolutionize chemistry as envisaged by some of the movement's leaders in the early 2000s, is essentially the same: it is the vagueness of its boundaries; the lack of clearly delineated features.

According to the standard narrative presented above, the entire green chemistry concept reposes on two major oppositions. Firstly, green chemistry rejects the old, unsustainable, and toxic 'brown' chemistry. Secondly, it is opposed to the failed 'end-of-pipe' regulatory solutions as developed in the 1970s and 1980s. Green chemistry aims to replace both these approaches. However, in both cases, the distinction between these supposedly contrasting philosophies becomes much less clear under closer inspection.

2.1. Brown vs Green Chemistry

The contrast between the 'brown' and 'green' chemistry[13] was explicitly formulated in the early 2000s by the political scientist Edward Woodhouse involved in the green chemistry movement in its early days (Table 4). It was then used as an example by the pioneers of green chemistry, James Clark and Duncan

---

12 Roberts, "Creating Green Chemistry," 137.
13 It is interesting to point out that chemistry is often given some colour. For example, an influential French chemist Guy Ourisson used the terms red chemistry and black chemistry to describe chemical products and processes responsible for accidents and pollution respectively. Guy Ourisson, "Chimie polluante, chimie non polluante, chimie dépolluante," in *La chimie, Université de tous les savoirs*, volume 18 (Paris: Odile Jacob, 2002): 219–227.

Macquarrie, in one of the first textbooks of green chemistry in 2002,[14] and then republished by Woodhouse in a slightly revised version in 2005.[15] In various forms, this differentiation continued to circulate as a metaphor well into the 2010s influencing subsequent generations of chemists.[16]

Table 4: Brown vs Green Chemistry (Woodhouse and Breyman, 2005 version).

Brown chemistry (20th century)	Green chemistry (21st century)
• Start with a petroleum-based feedstock; • dissolve it; • add a reagent; • react the compounds to produce intermediate chemicals; • put these through a long series of additional reactions • to yield megaton quantities • of potentially dangerous final products; • release these into ecosystems and human environments without knowledge of long-term effects, • without going through gradual scale-up to learn from experience; and • in the process create millions of tons of hazardous wastes as by-products.	• Design each new molecule so as to accelerate both excretion from living organisms and biodegradation in ecosystems; • create the chemical from a carbohydrate (sugar/starch/cellulose) or oleic (oily/fatty) feedstock; • and rely on a catalyst, often biological, • in a small-scale process • that uses no solvents or benign ones • and requires only a few steps, • creating little or no hazardous waste as by-products; • to yield small quantities of the new chemical for exhaustive toxicology and other testing, • followed by very gradual scale-up and learning by doing.

The distinction between brown and green chemistries boils down to two statements: 1) green chemistry is a new trend aiming to replace the old brown chemistry and 2) green chemistry cares for people, resources, and the environment, unlike its predecessor. Both Woodhouse and Breyman in the 2005 article and Clark and Macquarrie in the 2002 textbook explain that the comparison is exaggerated for the purpose of the argument. Still, the contrast permeates a wide range of publications and this framing, even if purposefully simplified, is deeply

---

14 James H. Clark and Duncan Macquarrie, *Handbook of Green Chemistry and Technology* (Oxford: Blackwell Science, 2002), 2.
15 Woodhouse and Breyman: "Green Chemistry as Social Movement?," 200–201.
16 See for example: Avtar Matheru and Pascale Champagne, "Brown to green and sustainable chemistry" in *Current Opinion in Green and Sustainable Chemistry*, 2, 2016, iii–iv; Sarah A. Green, "Progress and Barriers," in Mark A. Benvenuto: *Sustainable Green Chemistry*. Berlin: Walter de Gruyter, 2017, 17–28.

problematic. In fact, even Anastas in his recent paper (2018) "Origins and Early History of Green Chemistry" appears to distance himself from this simplistic reading, to which he contributed in the 1990s.[17] The problem lies in the fact that if we get rid of this contrast between the two chemistries, it deals an important blow to the entire green chemistry project, since this division is constitutive of green chemistry's identity. In practice, the brown/green chemistry divide is faulty for two main reasons: historical and epistemological.

First of all, this contrast is questionable because it gives the impression that there were no environmentally friendly chemistries before the rise of green chemistry and that the chemists of yore were unaware of the environmental impact of their activities. This assumption was rejected by Anastas himself in 2018. In his paper, he mentions traditional reference points such as chemical catastrophes and Carson's *Silent Spring*, but in the section "Prehistory of Green Chemistry," he introduces a few essentially novel elements previously lacking in the core narrative. He notably discusses environmental regulations preceding Rachel Carson's book, such as the Water Pollution Control Act (1948) and California's air pollution regulations (1946). This alone constitutes a departure from a somewhat naive interpretation widespread in older green chemistry stories in which Rachel Carson and her contemporaries would have suddenly discovered that the chemical industry was polluting, and the pre-1960s years had constituted a period of unfaltering faith in scientific progress.

Yet Anastas goes even further back in time. He talks about the inventor of catalysis, a crucial tool in green chemistry, Jons Jacob Berzelius, and the inventor of ionic liquids, another hot topic in the field, Paul Walden. Of course, these scientific pioneers, as important as they were, did not work with any environmental concerns in mind. They were chemists, not green chemists, whose contributions were nevertheless useful for the field. But then, Anastas devotes a paragraph to another chemist: Alfred Nobel. This time, he insists on Nobel's agency: it was Nobel's explicit intention to make explosives safer. Thanks to his efforts, there were fewer tragic accidents and risks associated with certain chemical substances became smaller.

Anastas links Nobel's approach to the philosophy of a more contemporary chemist, Trevor Kletz, who advocated in favour of the concept of "inherent process safety" from the 1950s. Anastas writes:

---

17   Anastas, "Origins and Early History of Green Chemistry."

In identifying key principles for inherent safety — smaller amounts of hazardous material in inventory, substituting hazardous material for nontoxic and nonflammable material, only using hazardous material in less hazardous form and conditions, and simplifying systems so they are easier to understand — he made it easier for plants to adopt an inherently safer approach. One of Kletz's main arguments, and the title of one of his papers, was "What you don't have, can't leak." Similar principles of inherently safer design are progenitors of the principles of green chemistry.[18]

Martyn Poliakoff, another very influential green chemistry scholar, called Kletz a "new father figure" of the entire green chemistry movement (suggesting he discovered his works after the green chemistry concept had been established).[19] The fact that Kletz was not cited in the early green chemistry works may be a result of the fact he was British, therefore less known to Anastas and the American EPA scholars, and worked in the industry and not in academia. What matters is that laboratory safety corresponds to principle 12 of green chemistry and the idea of prevention lies at the heart of the entire philosophy of green chemistry practitioners. Kletz's concept is consequently largely in line with what Anastas and Warner developed in the 1990s.

The fact that green chemistry had forerunners is neither surprising nor revealing, even though some scholars, such as the industrial chemist Mark A. Murphy, argue that the way the history of green chemistry has been written to date tends to obfuscate the importance of the pre-1990s pioneers.[20] What I insist here on is the fact that if it is possible to trace the roots of the field to Trevor Kletz or even Alfred Nobel, the chronological division between the old 'brown' and the new 'green' chemistry does not stand scrutiny. There have always been chemists trying to make things safer and address dangers posed by the chemical industry. In fact, conflicts and civil litigations concerning the safety of chemical installations, pollution, and toxic wastes are as old as the industry itself and the relevant regulatory framework developed organically since the nineteenth century; it is one of the most established topics in the history of chemistry.[21] If so,

---

18 Anastas, "Origins and Early History of Green Chemistry," 6.
19 Martyn Poliakoff, "Kletz T. A new father figure?," *TCE: The Chemical Engineer*, No. 866 (2013): 42–37.
20 Mark A. Murphy, "Early Industrial Roots of Green Chemistry and the history of the BHC Ibuprofen process invention and its Quality connection," *Foundations of Chemistry* 20 (2018): 121–165.
21 Geneviève Massard-Guilbaud, *Histoire de la pollution industrielle - France, 1789–1914* (Paris: EHESS, 2010); Nathalie Jas and Soraya Boudia, *Toxicants, Health and Regulation since 1945* (Cambridge: Cambridge U. Press, 2013); Bradley D. Snow, *Living with Lead: An Environmental History of Idaho's Coeur D'Alenes, 1885–2011*

green chemistry makes a part of a much older trajectory. Greenness can be associated with the chemistry of good intentions: the pioneers of green chemistry are those who tried to make things safer, less toxic, less pollutant, and so on. As such, the historical shift from brown to green has to be deeply nuanced putting the supposed newness of the green chemistry aspirations in question.

One could retort that what is different about green chemistry, at least as seen by Woodhouse, are the starting materials; the ambition was to replace non-renewable materials with biomass-derived ones. But even so, biomass studies proliferated since the 1970s (so long before the supposed emergence of green chemistry), and the history of chemurgy (industrial use of crops and other organic raw materials) and especially of wood and resin chemistry stretches much further in time.[22] There is no doubt we are still far from replacing petroleum as the blood of our economy, but one cannot claim that such non-petroleum chemistry is something new.

So perhaps the difference between the two chemistries, brown and green, is then purely epistemological in nature? Perhaps greenness and brownness have always coexisted in the chemical industry and the balance simply shifted from one to another? Perhaps we can place Nobel, Kletz, and chemurgists on the right side of history, and call them the good guys, as opposed to the chemists and chemical industries uncritically exploiting fossil fuels and rejecting carelessly toxic wastes into the environment?

Here Anastas makes another interesting observation in his recent history article:

> No company was trying to create toxic waste or to pollute cities. The aim has always been to help solve problems, not create them. Dichlorodiphenyltrichloroethane (DDT), for instance, was first developed and used to protect crops and kill disease carrying insects such as lice and mosquitos. It was considered a miracle pesticide, saving countless lives from typhoid and malaria. But with such a narrow focus on the positive functions of DDT, the negative consequences of its use went largely unknown until evidence began to surface of DDT's ill effects.[23]

(Pittsburgh: University of Pittsburgh Press, 2017); Ernst Homburg and Elisabeth Vaupel, *Hazardous Chemicals: Agents of Risk and Change, 1800–2000* (New York: Berghahn Books, 2019).

22 Mark R. Finlay, "Old Efforts at New Uses: A Brief History of Chemurgy and the American Search for Biobased Materials," *Journal of Industrial Ecology* 7 (2003): 33–43; Marcin Krasnodębski, "Intitut du Pin et la chimie des résines en Aquitaine (1900–1970)" (PhD diss., University of Bordeaux, 2016).

23 Anastas, "Origins and Early History of Green Chemistry," 2.

Whereas many claim to practice green chemistry, no one considers themselves to be a 'brown chemist.' Brown chemistry is an *a posteriori* reconstruction of a number of practices that are considered to be bad for the environment. There is nothing wrong with such re-reading, but, epistemologically speaking, brown and green chemistry do not belong to the same category. Green chemistry is considered by its proponents to be a specific set of normative tools and practices to achieve their goals; it is a consciously chosen self-descriptor. Brown chemistry is a polemical and critical term used by the same people to describe the chemistry they reject. None of the 'brown chemists of yore' would embrace the label or purposefully conduct his or her activity to harm the environment and human health.

However, if the goal of chemistry has always been to solve problems and make things safer and healthier, is there any process or product that could not be considered green at the moment of its introduction? DDT and similar substances were believed to have saved thousands of lives and be safe for humans, does it make them green?

Some may consider the question unfair and claim that green chemistry, as a novel concept, offers new tools that allow us precisely to avoid the tragedy of DDT. And yet, by the end of 2000, almost ten years after the supposed birth of green chemistry, a paper published in the flagship journal *Green Chemistry* mentioned DDT in a warm way suggesting the necessity of re-evaluating its use.[24] It is interesting to see an opinion like this expressed in a journal devoted to a field tracing its roots to Rachel Carson, but the point is not to discuss here whether the article is right or not (I return to this paper and frontiers of greenness in chapter 4). The point is that the historical and philosophical break supposedly brought about by green chemistry and separating it from brown chemistry is nowhere to be found. On the contrary, we see large decades-long processes and rational re-evaluations of older solutions organically evolving in a world full of political, social, and economic constraints. The criteria of what is good and bad for humankind and the environment changed over time and some activities considered harmful today were cherished only a few decades ago. If the brown/green division is more of a dynamic process, there is a risk that some activities of early green chemists may become seen as brown in the future. If so, the descriptive and historical value of the differentiation loses much of its allure. Any best chemistry in a given moment becomes green in the eyes of its practitioners.

---

24 Neil Winterton, "Chlorine: the only green element – towards a wider acceptance of its role in natural cycles," *Green Chemistry* 2 (2000): 173–225.

All this argument is not to undermine the ambition of green chemistry (to reduce environmental impact, to use more renewables, etc.), but to reject the simplistic binary logic pervasive in the self-representation of the field. It is by deconstructing its underlying assumptions, as presented in the dominant narrative on the history of green chemistry, that we can understand its current state, and perhaps offer a more informed understanding of its dynamics and perspectives.

2.2. Green chemistry versus 'command and control' approach

If the brown/green division does not stand under scrutiny and fails as a foundation for the concept of green chemistry, what about the second crucial divide that stems from the official narrative? Green chemistry was supposed to break with the obsolete, cumbersome 'command and control' approach to environmental pollution that spurred a plethora of strict environmental regulations that the industry had to take into account. The standard story on green chemistry states that it was because this 'end-of-pipe' logic failed to solve the problems in the long run, that the emergence of green chemistry was necessary.

Jody Roberts captures the problem most skilfully in his dissertation, in which he shows a graph illustrating the exponential growth of environmental regulations weighing on the chemical industry in the US throughout the twentieth century. As he puts it:

> At the Green Chemistry Summer School held in the summer of 2004, this graph was omnipresent. ... Berkeley 'Buzz' Cue, a former VP at Pfizer and now working as an independent consultant to the chemical industry, told the participants gathered there that this would be the scariest slide they would see. This sentiment was repeated recently by Paul Anastas at the Chemical Heritage Foundation's Second Annual Innovation Day.[25]

Then Roberts cites Anastas and Warner's book from 1998:

> It is staggering to imagine that in many of the large chemical companies in the United States, expenditures on research and development are equal to expenditures on environmental health and safety. In this statement lies the illustration that the one of the true victims of the costs of using and generating hazardous substances is the further growth and innovation of the science and industry of chemistry. Universities and small colleges are meeting the challenge of the cost of waste disposal from chemistry labs, both educational and research, by reducing either the number of laboratories or reducing the scale upon which laboratory experiments are run.[26]

---

25 Roberts, "Creating Green Chemistry," 110.
26 Anastas and Warner, *Green Chemistry*, 31.

Roberts comments on this approach:

> Indeed, this was one of the primary motivations in moving away from the traditional 'command and control' approach to pollution and regulation, as the Pollution Prevention Act points out. The rising costs of regulatory compliance, the argument goes, strips away money that could be used for further research and development costs. It's important to note that the rhetoric here is not geared towards threats to individual companies and their burdens, or the chemical industry as a whole. Instead, the language being used suggests a threat to the very practice of chemistry. Chemistry, if it does not change, will fall apart due to the economic stresses it currently faces. The authors are careful to link the economic imperatives to the environmental and social issues to which they are necessarily connected.[27]

Roberts argues that green chemists believe to have created a win-win situation for business and the environment alike: green chemistry prevents pollution, so the industry does not need to suffer the costs of pollution regulation. Anastas and Warner underline in their 1998 book that green chemistry is not just a chemistry that is better for the environment, less polluting and safer, but that it is above all a chemistry that is good for business. In a certain way, profitability for the industry is the unspoken thirteenth principle of green chemistry that has permeated the entire field.

Again, what is significant is that the line was very clearly drawn between the old 'command and control' or 'end-of-pipe' approach, which was overly expensive, and the new green chemistry one bringing together scientific progress, environmental safety, and profitability.

This differentiation, so important for the identity of early green chemistry proponents, is, however, historically problematic. The molecular pollution prevention advocated by green chemists was not some radical departure from earlier environmental strategies, but the crowning of a decades-long development. One of the godfathers of green chemistry, Joe Breen, the first executive director of the Green Chemistry Institute, wrote himself in 1992 that the new paradigm of pollution prevention is the fruit of a process initiated 20 years before. According to him, first, there was the control of the "end of pipe" releases, then the more mature practical reflection on "waste management technology," which evolved into "waste minimization" from which there was only one step to "pollution prevention."[28] The Pollution Prevention Act from 1990, occasionally presented as

---

27 Roberts, "Creating Green Chemistry," 111.
28 Joseph J. Breen and Michael J. Dellarco, *Pollution Prevention in Industrial Processes: The Role of Process Analytical Chemistry* (Washington, DC: American Chemical Society, 1992), 2.

the foundational legislation that spurred green chemistry, was itself a result of previous activism, lobbying, and research. The pollution prevention philosophy was, indeed, widely practised by the chemical industry as early as the middle of the 1980s even if without necessarily having been called that way.[29]

This historical reconstruction begs also a more epistemological question. Isn't green chemistry merely a 'command and control' approach extended on the molecular level? The reason why the companies want to avoid pollution is because of the pre-existing environmental legislation, and because the legislation itself (e.g. Pollution Prevention Act) encourages this approach.

The point is to show that green chemistry places itself on a large continuum and is not some radically new approach that was developed from scratch in the 1990s. Green chemistry is a type of environmental pollution chemistry focused on pollution prevention that became simply somewhat more formalized in the 1990s. If so, why did its early promoters insist so much on the novelty of their approach? One may be led to a somewhat cynical interpretation according to which the core motivation of the old environmentally-minded regulation philosophy was to protect the environment against pollution even if it meant additional costs for the industry, while green chemistry was invented to protect the business against the costs of protecting the environment. In fact, the association between the greenness of American dollars and the greenness of nature is a common leitmotif in many descriptions of the American variant of green chemistry.[30] This is of course an unfair representation of a tremendous amount of genuinely positive work conducted by environmentally minded chemists who identified themselves with green chemistry over the last twenty years. The crude reductionism of green chemistry to the spin given to it by some of its early practitioners should not cast a shadow on the entire field. However, what is at stake is the purely theoretical framing of green chemistry as a new separate thing, different in some way from what used to be practised before. Just as much as with brown chemistry, the differentiation between green chemistry and the 'command and control' approach is not solidly grounded in the historical analysis. It was a tool for building a disciplinary identity, its sense of separateness, but in a broader scheme of things, this division was largely artificial.

---

29 Murphy, "Early Industrial Roots of Green Chemistry."
30 Laure Gilles and Sylvain Antoniotti, "Chimie durable et parfumerie," *Nez, la revue olfactive* (2020). https://mag.bynez.com/science/chimie-durable-et-parfumerie/; See also: Pietro Tundo, "Green Chemistry on the Rise," *Chemistry International* 29 (2007): 5–7.

Conclusions

The purpose of this short introductory chapter was to highlight the fact that there are preliminary problems with the core narrative and self-representation of green chemistry as a distinct field of study. While green chemists tell themselves a certain story on the origins of the field and explain how this story differentiates it from other philosophies ('brown chemistry', 'command and control' approach), a more careful analysis of their claims shows important gaps in their reasoning. These difficulties undermine the entire project and threaten the identity and separateness of green chemistry, reducing it to a mere buzzword without inherent meaning. If it is not possible to delineate the frontiers of green chemistry, concepts and practices incompatible with original intentions may substitute the elements that were considered initially essential or, worse, that there would be no criteria to exclude anything from this broad umbrella of ideas. In this sense, all chemistry can become green chemistry, as hoped by its promoters in the early 2000s, but not because of some genuine evolution of methods, but because the definition becomes overly vague and anyone can claim it.

To be sure, green chemistry practitioners can retort that the sole fact that some people abuse the name does not mean it is devoid of meaning and that real green chemists can very well identify what counts as their field of study and what does not. But this only creates another line of division: between the holders of the 'orthodoxy' who impose allowable margins of greenness and the outsiders, 'heretics,' deforming its true meaning. Then again, who has the right to define the boundaries? Those who conduct empirical studies in the laboratories and tag their results with green chemistry keywords? Those who teach at the universities? The editors of journals (but which ones?) who select articles? Those who invented the term? Or perhaps those who write its history today?

It may be argued though that this confusion could be avoided if we returned to the sources, to the original message, or to the 'holy texts' of the domain. Perhaps if we read what was originally considered green chemistry, we would be better equipped to evaluate greenness claims made by the others. In the following chapter, I delve into the early history of green chemistry in order to counter this line of argumentation.

Chapter 2: The formative 1990s

Throughout the entire year 2016, the *Green Chemistry* journal sumptuously celebrated the 25th anniversary of green chemistry with a series of articles, published every month, devoted to the twelve principles.[1] A quarter of a century is a significant timespan and the journal's editors wanted to summarize the most important achievements of green chemists as well as the evolution that the field had undergone. But why exactly twenty-five years? Paul Anastas regularly wrote in his own recollections that the term had been invented by him and his collaborators working in the EPA in 1991. It is one of the elements of the core narrative that did not undergo any significant changes and still in 2018 Anastas explained that it had been "[i]n 1991, when the term "green chemistry" was officially defined."[2] This date was a reference point in the publications from early on. One article from 2002 explains that:

> Shortly after the passage of the Pollution Prevention Act [1990], it was recognized that a variety of disciplines needed to be involved in source reduction. This recognition extended to chemists, the designers of molecular structures and transformations. In 1991, the Office of Pollution Prevention and Toxics in the U.S. Environmental Protection Agency launched the first research initiative of the Green Chemistry Program, the Alternative Synthetic Pathways research solicitation. Foundational work in chemistry and engineering at the National Science Foundation's program on Environmentally Benign Syntheses and Processes was launched in 1992, and formed a partnership with EPA through a Memorandum of Understanding that same year. In 1993, the EPA program officially adopted the name "U.S. Green Chemistry Program."[3]

However, when it comes to confirming the exact years and the authorship of concepts, there is only Anastas's testimony. As Jody Roberts put it already back in 2006: "Unfortunately, Anastas is one of very few people from those early days

---

1 See the introductory article published in the first issue in 2016: Paul Anastas, Buxing Han, Walter Leitner, Walter Leitner and Martyn Poliakoff, "'Happy silver anniversary:' Green Chemistry at 25," *Green Chemistry* 18 (2016): 12–13.
2 Paul Anastas, "Origins and Early History of Green Chemistry," in *Advanced Green Chemistry. Part 1: Greener Organic Reactions and Processes*, ed. I. T Horváth and M. Malacria (London: World Scientific, 2018): 1–17, 6.
3 Paul Anastas and Mary M. Kirchhoff, "Origins, Current Status, and Future Challenges of Green Chemistry," *Accounts of Chemical Research* 35 (2002): 686–694, 686–7.

at the EPA still living."[4] While there may be no reason to question the year itself (even though not a single EPA's publication uses the term green chemistry in 1991), one can be sceptical about the actual content of the concept in that period. Is the EPA's green chemistry in 1991 the same green chemistry as today or even similar to the green chemistry of the 12 principles published for the first time in 1998? How was the term understood by different stakeholders over the period? Did all people agree on its meaning?

The 1990s were a melting pot of ideas with many different green chemistries struggling for recognition. The purpose of this chapter is to provide a guide to the formative years of the concept and to understand why only one version of green chemistry emerged victorious. A word of clarification: in this chapter, I follow the name (green chemistry) and its uses. The early story of green chemistry becomes even more nuanced when we contrast it with other similar conceptualisations to which I return in chapter 7.[5]

1. Green chemistry outside the Anglosphere in the late 1980s and early 1990s

The first thing that needs to get out of the way, and that is regularly ignored in many publications on the history of green chemistry, is that the term itself had widely circulated in the non-English speaking countries long before it appeared in the EPA's publications. In 1987, two Italian scholars, Italo Pasquon and Luciano Zanderighi, published a book entitled *La chimica verde: le utilizzazioni dei prodotti vegetali e le biotecnologie* [Green chemistry: uses of plant products and biotechnologies].[6] It was an introductory textbook on the use of bio-sourced materials, such as ethanol from plants, to replace petroleum. The motivations of the authors were above all environmental as they deplored the unsustainable character of fossil fuels. According to Pasquon and Zanderighi, the time was ripe not only for systematic exploitation of plant resources but also for the development of synthesis processes that would imitate nature. They argued for the use of

---

4   Jody Roberts, "Creating Green Chemistry: Discursive Strategies of a Scientific Movement" (PhD diss., Faculty of Virginia Polytechnic Institute and State University, 2006), 70.
5   For a slightly different and more focused take on the topics discussed in this chapter see: Marcin Krasnodębski, "Lost green chemistries: history of forgotten environmental trajectories," *Centaurus* (accepted).
6   Italo Pasquon and Luciano Zanderighi, *La chimica verde: le utilizzazioni dei prodotti vegetali e le biotecnologie* (Milano: Hoepli, 1987).

crops' surplus to produce fuels, fertilizers, and all sorts of chemical substances. The Italian scholars insisted on the fact that the shift towards plant-based chemistry might be also environmentally friendly due to the biodegradability and low toxicity of such products. At the same time, they warned that not necessarily all plant-based products were less toxic, and that some agricultural processes could be highly polluting, and they did not ignore the threat of industrial crops to biodiversity. In other words, they explained that any shift from the petroleum-based industry to a biomass-based one would require careful planning and lots of research on avoiding pollution and toxicity.

One could argue that this way of defining green chemistry is much narrower than the EPA's green chemistry as explained in the twelve principles from 1998, therefore the two concepts, Italian and American, cannot be really compared. In fact, only principle 7 of the American principles refers to renewable sources, whereas for the Italian chemists the question of biomass was the top priority and all the rest was built around it. Not to mention that, unlike the EPA scholars, the Italians failed to produce any disciplinary framework or journals to popularize their concept. The bottom line would be that the Italian approach prefigured in a somewhat simplistic way what would become green chemistry in the US in the following years. And yet there are two problems with this line of argumentation.

Firstly, if one takes as a reference not exclusively the 12 principles but the brown/green contrast, as developed by Woodhouse and then embraced by Clark, the Italian position seems to be more in line with the modern interpretation of green chemistry. In fact, as I show in the next chapters, principle 7 trumped many others and the renewability of raw materials became a more and more important component of greenness over the years. In a certain way, Pasquon and Zanderighi had better predicted in the 1980s the major trends in green chemistry of the 2020s than Anastas and Warner did in the 1990s.

Secondly, while "chimica verde" was certainly not as successful as the American green chemistry, it was not ignored either. In Italy, the concept was, for example, presented in publications concerning industrial pollution control.[7] Pasquon, a prominent professor of chemical engineering who specialized in organic chemistry and polymers,[8] recommended his book in the classes at the

---

7   Giovanni Bianucci and Esther Ribaldone Bianucci, *Il trattamento delle acque residue industriali e agricole* (Milano: Hoepli, 1992), 62–64.
8   Article in the national Italian encyclopaedia Treccani, accessed 18/02/2022, https://www.treccani.it/enciclopedia/italo-pasquon.

Polytechnic University of Milan in the 1990s.[9] The term chimica verde also circulated in the Italian chemical industry. It was also in Italy where the first known industrial project named green chemistry (on biodegradable plastics) was funded by the chemical company Ferruzzi. The project generated enough interest abroad to be noticed in a publication in *New Scientist* in 1989.[10] This was probably the first time the term green chemistry was used in English in a sense close to the contemporary one. While there is probably little connection between Ferruzzi's project and the work of Pasquon and Zanderighi, the point is that the term began circulating in Italy over the period and at least some Italian chemists had been familiar with it before the EPA started promoting its own green chemistry. This seems to be confirmed by the fact that in the early EPA's publications on green chemistry, Italians were among the most active non-American scholars. In the 1996 book on green chemistry edited by Paul Anastas, in which he offered some of the first definitions of the concept in the introductory chapter, five out of seventeen chapters were co-written by Italian contributors (and four by Italians from Milan-based institutions).[11]

In addition to Italy, the term green chemistry was used to describe some form of plant- and biomass-based chemistry also in France. In 1990, a booklet entitled *La chimie verte: Quelles stratégies pour les industries du sucre et de l'amidon* [Green chemistry: what strategy for sugar and starch industries] was published by an INRA researcher (Institut National des Recherches Agronomiques [National Institute of Agricultural Research], one of the main state-owned research organizations in France).[12] Two years later, one French pharmacist published the book *Chimie verte: de la plante au médicament* [Green chemistry: from plant to medicine] opting for a somewhat narrower interpretation of the term in the context of pharmaceutical uses.[13] Both texts assume nevertheless that green chemistry is a type of biomass chemistry. The term was also interpreted this way in the INRA's publication and communications for many years. While these texts

---

9   The teaching curriculum of the Milan Polytechnic in 1996–1996, accessed 18/02/2022, https://www.ingindinf.polimi.it/fileadmin/user_upload/scuola/programmi_insegnamenti_veccio_ordinamento/1995-96.pdf, 57.
10 Debora Mackenzie, "Technology: Italian firm first with 'truly biodegradable' plastic," *New Scientist* (25 November 1989).
11 Paul Anastas and Tracy C. Williamson, ed., *Green Chemistry. Designing Chemistry for the Environment* (Washington D.C.: American Chemical Society, 1996).
12 Arnaud Malerbe, *La chimie verte: Quelles stratégies pour les industries du sucre et de l'amidon* (Grignon: INRA, 1990).
13 Philippe Jaussaud, *Chimie verte: de la plante au médicament* (Paris: SUTIP SA, 1992).

did not explicitly rejected the petroleum paradigm, the focus on renewability and more environmentally friendly agricultural chemistry was certainly there. I do not discuss the French case in any more detail here, because I return to it in chapter 6, but it should be noted for now that in France this separate green chemistry tradition persists until this day.

Finally, the repertoire of international green chemistry forerunners would not be complete without one more article that has been occasionally mentioned in the publications on the history of green chemistry: Pavel Drašar's "Zelená Chemie: Sen nebo Realita (Minimum Impact Chemistry)" [Green Chemistry: Dream or Reality (Minimum Impact Chemistry)] published in the Czechoslovakian *Chemicke Listy* [*Chemical Letters*] in 1991.[14] Interestingly, the original title is in Czech but not the parenthesis. Minimum Impact Chemistry was apparently Drašar's choice to describe his philosophy in English. A very short English abstract states that "The article brings some ideas for provoking at least discussion on the improvement of the practices of chemistry. GLP/GMP is used as the practical example of possible regulation which will also mean a useful tool in the future negotiations of chemists and environmentalists." GLP (Good Laboratory Practices) and GMP (Good Manufacturing Practices) are at the heart of Drašars argument and it is through their extension that he envisions the greening of chemistry. In the body of the text, he notably calls for the development of GGLP – Good and Green Laboratory Practices.

The article itself, while short and rather general, indicates an interesting tension. On the one hand, Drašar is concerned about the rise of green political movements that often perceive the chemical industry as their main enemy. He notes, for example, the electoral breakthroughs of environmentalists in Austria. On the other hand, he admits he is "terrified" of the general level of environmental pollution and lack of management thereof. The concerns about both the environment and the public image of chemistry indicate that Drašar's mindset was very similar to the one of the American pioneers of green chemistry in the 1990s. The article did not give rise to any distinct Czech philosophy on environmental problems but was rather emblematic of the general *Zeitgeist*. Unlike some other international forerunners, however, Drašar's work has been at least somewhat preserved in the collective memory, and the article has been notably cited by one of the most influential green chemistry theorists, Roger Sheldon.[15]

---

14 Pavel Drašar, "Zelená Chemie: Sen nebo Realita (Minimum Impact Chemistry)," *Chemicke Listy* 85 (1991): 1144–1149.
15 Roger Sheldon, "Green chemistry and resource efficiency: towards a green economy," *Green Chemistry* 18 (2016): 3180–3183.

Overall, whereas the Italian, French, and Czech ways of understanding green chemistry were not identical, they all presented a range of problems that would be at the heart of the green chemistry project in the following years. This is not to say that Anastas did not 'invent' the term on his own in 1991 within the walls of the EPA. However, there is no doubt that this type of terminology was popular in the period in many European countries. Therefore, placing the birth of green chemistry in 1991, a year from which there is no single publication on the topic coming from Anastas or his colleagues, appears to be problematic at best.

2. Anastas's green chemistry in the US before 1998
2.1. Preliminary remarks

In the dominant narrative on the history of green chemistry, the first major event in the field was a symposium held in Chicago in 1993 sponsored by the Division of Environmental Chemistry of the American Chemical Society (ACS). One year later, in 1994, its papers were gathered and published in the book *Benign by Design: Alternative Synthetic Design for Pollution Prevention*.[16] The standard narrative indicates that more books under the auspices of the ACS followed in 1996 and a series of the Gordon Research Conferences (GRC) featuring Anastas and his EPA colleagues took place in 1996 and 1997. After these early events, Anastas and Warner finally published the famous 1998 book *Green Chemistry: Theory and Practice*, which summarized the key precepts of the new field, expressed elegantly through the 12 principles.

The typical way the pre-1998 story is presented has some problems. For example, Jody Roberts found out in his investigation that the Gordon Research conferences from 1996 and 1997 had been filed under the name "Green Chemistry" in the GRC on-line database, but originally they were in fact entitled "Environmentally Benign Organic Synthesis" and the term green chemistry did not appear at all on the program in 1996, and in 1997 it was mentioned only twice, including one paper by Anastas on "Principles of Green Chemistry."[17] Anastas explains in his history paper from 2018 that "the Gordon Research Conference

---

16 Paul Anastas and Carol A. Farris, ed., *Benign by Design: Alternative Synthetic Design for Pollution Prevention* (Washington, D.C.: ACS Symposium Series, 1994).
17 Today, the database includes the original name "Environmentally Benign Organic Synthesis," but both conferences make part of the green chemistry series. See for example the website of the 1996 conference, accessed 26/02/2022, https://www.grc.org/environmentally-benign-organic-synthesis-conference/1996/.

Foundation objected to naming the original conference 'Green Chemistry' as it was viewed as too 'trendy.'"[18] There is no reason to believe otherwise, but the problem lies elsewhere. It is about retroactively attributing to publications, seminars, and conferences the green chemistry label. If one starts labelling as green chemistry research that had been conducted before the term was conceived and popularized, this begs the question where to stop, blurring the boundaries of the concept even further. Did all participants of the conference adhere to the principles of green chemistry? Did they even know that what they practised was green chemistry?

Overall, what matters is that while the pre-1998 history is mentioned in some publications on the origins of green chemistry, it is the 1998 book by Anastas and Warner that is always presented as the foundational work of the field. Commentators, be they historians or chemists, usually focused on its analysis to unravel green chemistry through the book's careful exegesis. It is nevertheless important to note that the 1998 work is considered to be a reference point because of its accessible language, especially the inclusion of the 12 principles, not because of the novelty or refinement of the concept as such. In fact, the pre-1998 publications by Anastas offer us many insights into the development of ideas that would transform into the 12 principles, as well as into the shifts in the way the EPA scholars approached the topic. It is important to reiterate: according to the standard narrative, green chemistry had been born in 1991, so seven years before Anastas and Warner published their seminal work. As such, their book was, above all, an attempt to reach a wider public, but it was based on the already existing theoretical structure. For that reason, I prefer to focus on the earlier publications because their content usually escapes commentators and because they summarize the ambitions of green chemistry in a more succinct way. In practice, I chose to present and discuss here the key fragments of the introductory chapters to two post-conference publications edited by Anastas in 1994 and in 1996, in which he lays down his own ideas about environmentally friendly thinking in chemistry. Both books are based on the symposia organized under the auspices of the Division of Environmental Chemistry of the American Chemical Society in 1993 and 1994 respectively. My argument is that the core of the green chemistry philosophy was formulated in these chapters and that they express much better the key features of what was actually practised as green chemistry in the late 1990s than the 1998 book.

---

18 Anastas, "Origins and Early History of Green Chemistry," 12.

2.2. *Benign by Design* (1994 book)

The first thing that has to be mentioned about the 1994 book (a follow-up to a conference held in 1993) is that, just as in the case of the Gordon Research Conferences, the term green chemistry was virtually absent in its pages.[19] Anastas in an interview with Roberts in 2004 explained this absence in the following way: "That's because we were focusing on the synthesis part of Green Chemistry. The Benign by Design was just a heck of a good book title."[20] This can explain the choice of the title, but not the fact that the name green chemistry was mentioned only three times in the entire book, and not a single time in any meaningful programmatic way. On the contrary, in his theoretical chapter "Benign by Design Chemistry" Anastas focused exclusively on what benign by design means, without using the term green chemistry. At the same time, none of the articles in the professional press published after the 1993 seminar, on which the book was based, mentioned the name even though the adjective "green" was already there (*Chemical Week*: "Chemists map greener synthesis pathways" and *Chemical & Engineering News*: "'Green' technology presents challenge to chemists").[21] This casts again a shadow on the idea that the term green chemistry was coined in a more or less formalized form already back in 1991.

This problem is directly related to another leitmotif recurrent in many histories of green chemistry. We often read that originally there were many competing names to describe this new philosophy: benign by design chemistry, environmentally benign chemical synthesis, benign chemistry, clean chemistry, sustainable chemistry, etc. This plurality of terminologies was notably mentioned by Paul Anastas and his collaborator Tracy Williamson already back in 1996, but more recently historians Jody Roberts and J. A. Linthorst conducted, in 2006 and 2010 respectively, scientometric cross-comparisons between these various terms.[22] Roberts found out, for example, that before a staggering rise of

---

19 Anastas and Farris, *Benign by Design*.
20 Roberts, "Creating Green Chemistry," 70.
21 David Rotman, "Chemists map greener synthesis pathways," *Chemical Week* 153 (1993): 56–57; Deborah L. Illman, "'Green' technology presents challenge to chemists," *Chemical & Engineering News* 71 (1993): 26–30.
22 Paul Anastas and Tracy C. Williamson, "Green Chemistry: An Overview," in *Green Chemistry. Designing Chemistry for the Environment*, ed. Paul Anastas and Tracy C. Williamson (Washington D.C.: American Chemical Society, 1996), 1–17; Roberts, "Creating Green Chemistry," 72; J. A. Linthorst, "An overview: origins and development of green chemistry," *Foundations of Chemistry* 12 (2010): 55–68, 56.

the popularity of the term "green chemistry" in 1999 and 2000, "environmentally benign chemical synthesis" was more popular. Linthorst did not search for this one, but his results show that terms such as "benign chemistry" and "clean chemistry" were barely ever used. A quick scientometric search in Scopus only confirms that these terms are not really comparable to what green chemistry has become and that at no point "clean chemistry" or "benign by design," or even "environmentally benign chemical synthesis" were used beyond a narrow number of publications connected to the EPA scholars in the 1990s. More importantly, all scientometric analyses show that the name "green chemistry" was also barely used before 1996 and that it was not until 1998 when it really took off.

Still, if "benign by design" was a sister concept that became later a blueprint for green chemistry in the EPA, it would be interesting to learn a bit more about it. How does Anastas define it in 1994? The preface of the book states

> Benign by design chemistry is synthetic chemistry designed to use and generate fewer hazardous substances. The ultimate goal of benign chemistry research is to develop and institute alternative syntheses for important industrial chemicals in order to prevent environmental pollution.[23]

In the book's introductory chapter (entitled "Benign by Design Chemistry"), Anastas describes the philosophy of pollution prevention as practised in the 1980s in the US, highlighting its insufficiencies:

> When the concept of pollution prevention first was introduced as an approach to environmental problems, the majority of its early manifestations were in the form of housekeeping solutions. Reducing leakages in piping systems, covering vats and vessels which hold volatile substances to reduce evaporation, reducing loss of material through overspray in spraying applications ... The potential for pollution prevention, however, is far more fundamental. It requires a change in the manner in which products and processes are designed from their inception. Without question, all approaches toward preventing pollution should be assessed and implemented wherever possible. This chapter, however, will focus on the earliest design phase of chemical manufacture, the design of the synthetic sequence.[24]

So far the benign by design chemistry appears to be simply the pollution prevention philosophy, which had existed already for some time, applied to chemical synthesis. Then, however, a more precise definition comes along. This is, perhaps,

---

23 Anastas and Farris, *Benign by Design*, IX.
24 Paul Anastas, "Benign by Design Chemistry," in *Benign by Design: Alternative Synthetic Design for Pollution Prevention*, ed. Paul Anastas and Carol A. Farris (Washington, D.C.: ACS Symposium Series, 1994), 2–22.

the first published conceptualisation of what would become green chemistry in the following years:

> Benign By Design chemistry defines synthetic elegance on the basis of three factors:
> - Efficiency of synthetic methodology
> - Economically viable
> - Environmentally Benign
>
> The concept of synthetic efficiency has been discussed in terms such as carbon economy and atom economy … Both of these concepts poignantly illustrate the desirability of the incorporation of all of the atoms used in the transformation of the starting materials into the product. …
>
> Economic viability, simply stated is a pass/fail test for commercialization of a process to manufacture chemical products. If the synthetic technology cannot survive economically, the other virtues of the method quickly become irrelevant. It must be remembered, however, that the economic analysis must make sure to take into consideration all costs related to the manufacture such as those listed above (e.g., waste disposal, regulatory compliance, etc.) …
>
> [Concerning environmentally benign aspect] one can no longer consider just the yield and cost of feedstocks when selecting a synthetic route to the manufacture of a chemical product. Synthetic chemists and decision makers need to ask the following questions:
>
> What are the Toxicity Impacts of the Manufacture to Humans? This analysis must include all substances related to the synthesis. Toxic endpoints should include not only lethality (LD_{50}), but also endpoints such as neurological disorders, reproductive and developmental effects, etc.
>
> What is the Impact on the Living Environment? Considerations of direct toxicity to various plant life and wildlife should be included whenever possible.
>
> What is the Impact on the Larger Environment? The "larger environment" would include the effects such as stratospheric ozone depletion, atmospheric ozone generation, greenhouse gas generation, acidification and/or deoxygenation of lakes and streams, etc.
>
> Will This Increase the Potential for a Chemical Accident? While very often pollution prevention and accident prevention work hand-in-hand to minimize the risk to human health and the environment, this is not always the case and cannot be assumed without an analysis.[25]

To sum up the definition above, human toxicity, big environmental challenges, and accident risk are three criteria for qualifying something as environmentally benign. The environmentally benign processes and reactions would not gain popularity though, Anastas explains, if they were not also economically viable. Additionally, they should also aim for synthetic efficiency, which is seen as both economically and environmentally desirable.

---

25 Anastas, "Benign by Design Chemistry," 9.

Anastas enumerates later substances that should be accounted for in the process of greening: feedstocks, reagents, reaction media, by-products and impurities, catalysts, separation solvents, and distillation solvents. The rest of the chapter describes practices and strategies involving academia, the industry, and the government that can foster the development of benign by design chemistry.

The 1994 chapter demonstrates a remarkable capacity for synthesis and framing of different issues that Anastas used in many of his later publications and for which he became famous. Much of the conceptual framework presented here was later incorporated into the green chemistry publications, even if the term green chemistry was still conspicuously absent at this stage. This is not to say that the benign by design chemistry as presented in this chapter did not undergo any changes when it transformed into green chemistry properly speaking.

2.3. Green Chemistry. Designing Chemistry for the Environment (1996 book)

Two years after the benign by design book, in 1996, another major postconference publication (the seminar took place in 1994), edited by Paul Anastas and his collaborator from the EPA Tracy Williamson, was published under the title *Green Chemistry. Designing Chemistry for the Environment*. In this publication, the term green chemistry is finally omnipresent. The preface starts by explaining that

> Green chemistry focuses on the design, manufacture, and use of chemicals and chemical processes that have little or no pollution potential or environmental risk and are both economically and technologically feasible. The principles of green chemistry can be applied to all areas of chemistry including synthesis, catalysis, reaction conditions, separations, analysis, and monitoring.[26]

There is a clear parallel between this definition and the 1994 one. It is an attempt to broaden considerations and present green chemistry as an overarching concept going beyond chemical synthesis. Anastas and Williamson explain in the chapter "Green Chemistry: An Overview" that:

> Over the past few years, the chemistry community has been mobilized to develop new chemistries that are less hazardous to human health and the environment. This new approach has received extensive attention ... and goes by many names including Green Chemistry, Environmentally Benign Chemistry, Clean Chemistry, Atom Economy and Benign By Design Chemistry. Under all of these different designations there is a

---

26 Anastas and Williamson, *Green Chemistry. Designing Chemistry for the Environment*, XI.

movement toward pursuing chemistry with the knowledge that the consequences of chemistry do not stop with the properties of the target molecule or the efficacy of a particular reagent.[27]

Later they continue:

> Simply stated, Green Chemistry is the use of chemistry techniques and methodologies that reduce or eliminate the use or generation of feedstocks, products, by-products, solvents, reagents, etc., that are hazardous to human health or the environment.

They also repeat one of the main arguments of the 1994 book, namely the focus on the economic dimension of the methods they advocate: green chemistry is supposed to save money because less would be spent on clearing waste. Here Anastas and Williamson offer another revealing analogy on the character of the concept:

> Many in industry respond that the cost of complying with environment regulations in the United States is an environmental tax. If this is true, then Green Chemistry can be understood as a tax break waiting to be taken advantage of by industry.[28]

This is something I have already shown in chapter 1; in the case of green chemistry, greenness of the natural environment and of American dollars go hand in hand.

A very important statement comes next:

> It is important that chemists develop new Green Chemistry options even on an incremental basis. While all elements of the lifecycle of a new chemical or process may not be environmentally benign, it is nonetheless important to improve those stages where improvements can be made. The next phase of an investigation can then focus on the elements of the lifecycle that are still in need of improvement. Even though a new Green Chemistry methodology does not solve at once every problem associated with the lifecycle of a particular chemical or process, the advances that it does make are nonetheless very important.

This can be understood in the way that green chemistry is the methodology of "greening" chemistry. If so, there is no inherently green chemistry. In fact, Anastas and Williamson underline a few times that

> nothing is benign. All substances and all activity have some impact just by their being. What is being discussed when the term benign by design or environmentally benign chemistry is used is simply an ideal.[29]

---

27 Anastas and Williamson, "Green Chemistry: An Overview," 1.
28 Anastas and Williamson, "Green Chemistry: An Overview," 6.
29 Anastas and Williamson, "Green Chemistry: An Overview," 2.

In other words, Anastas and Williamson are toning down expectations: green chemistry is not here to revolutionize things but to show the direction of the coming evolution.

How to measure the overall increase in greenness?

> One must determine whether the new chemistry results in a net improvement to human health and the environment or whether additional incremental improvements are necessary in order to complement the original change in order to ensure that the entire synthetic pathway is more benign.[30]

To facilitate the task

> the work being done [in green chemistry] can be categorized according to what specifically is new or different about a green chemical or process in comparison to the conventional method. A logical breakdown of the chemical reaction results in four basic components: 1. Nature of the Feedstocks or Starting Materials 2. Nature of the Reagents or Transformations 3. Nature of the Reaction Conditions 4. Nature of the Final Product or Target Molecule.

A huge part of the chapter explores these four different categories in detail and provides a state of the art in the field.

Overall, the 1996 chapter offered a well-articulated conceptual backbone for green chemistry that would be almost entirely incorporated in the 1998 book by Anastas and Warner. Even more, I argue that this chapter actually captured better the core character of the green chemistry project in the late 1990s and early 2000s simply because the 1998 book expanded green chemistry in new directions that had not really been a part of the movement's initial identity. All in all, it only shows that green chemistry, even in the way Anastas understood it, was a dynamically evolving framework.

2.4. Laying down the foundations of green chemistry

When reading these two introductory chapters (from 1994 and 1996) that would become the foundations of one of the most exciting chemical concepts of the twenty-first century, we see a range of interesting phenomena. First of all, Anastas often defines the same terms multiple times. All the definitions are similar, sometimes complementary, but nevertheless leaving a lot of space for interpretation. Roberts identified in the 1998 work *Green Chemistry: Theory and Practices* no less than 9 statements trying to define green chemistry. In the 1994 and 1996 books, one could say that the entire introductory chapters are long composite

---

30 Anastas and Williamson, "Green Chemistry: An Overview," 6.

definitions trying to capture the meaning of the endeavour. This is not bad *per se*. Anastas was sometimes perhaps overly descriptive, but he was trying to flesh out his ideas and construct something original. At the same time, this lack of precision and constant reframing of the concept make it harder to pin it down and clearly draw frontiers of the supposed discipline; a problem that haunts green chemistry until today.

More interestingly, while in theory the 1996 green chemistry concept was supposed to be broader than the 1994 benign by design one, in practice the definitions were very similar. One could simply say that green chemistry was benign by design chemistry after rebranding. The overall ambition aside, little was added in the 1996 chapter. On the contrary, some elements were actually subtracted. The 1994 chapter advocated the evaluation of broader environmental impacts (ozone layer, global warming, etc.), but this dimension was completely absent in 1996. Furthermore, while both chapters mention alternative feedstocks, the 1994 one talks about the bio-based feedstocks and their potential to replace non-renewable fossil ones, but the 1996 one is exclusively concerned with reducing the feedstock's toxicity. Finally, in 1994 Anastas clearly posits his reflection as a part of the discipline called environmental chemistry, but he does not mention the term at all in 1996. It almost feels as if Anastas backed down on some of his previous ambitions, perhaps to make the idea of green chemistry more marketable to the business public. If we were to take "benign by design" and "green chemistry" definitions at face value from the chapters presented above, it is the former that appears to be larger and more sustainability- and environment-oriented, whereas the latter is its more modest and business-friendly sibling.

This is directly related to the problem that I have already mentioned in chapter 1: Anastas's focus on economic viability. When he calls green chemistry a "*tax break*," it may be a perfectly justified rhetorical turn to sell green chemistry to the industry. Anastas believed that any change in the chemical industry required convincing the private sector that being 'green' was good for business. It is, however, never clear in these texts where the frontier between the idea-pitching and the genuine concept-building is located. Should academic green chemistry scientists also think in terms of profitability? And what if the chemistry they propose is actually more expensive, but has tremendous positive environmental effects? In a way, this focus on not making things more expensive for the companies implementing green solutions is another feature that marked the entire history of the field. What would happen if all the effort of green chemists was directed into radically altering chemistry instead, and the costs of proposed solutions were evaluated not inside individual companies, but on the global society level?

This issue was also raised by Roberts:

> Green chemistry works within the current [economic] framework, assuming that the win-win situation can be created without undergoing any major changes.

He criticized the movement's myopia to the underlying socio-economic problems responsible for pollution and chemical disasters.[31] Roberts referred mostly to his own experiences from the 2004 Summer School, but there is no doubt that the origins of this line of thinking are deeply rooted in Anastas's publications from the 1990s.

3. Non-Anastas American green chemistries

Whatever we think about the 1994 and 1996 definitions of green chemistry, no other scholar in the 1990s wrote in a way as systematic and consistent about green chemistry as Paul Anastas. He was the figure of reference whose authority was growing over the years and dominated the movement precisely because of his talent for laying down broad conceptual frameworks. However, he was not the only one that had a specific vision on what green chemistry should become. On the contrary, some of his close collaborators had similar but nevertheless diverging perspectives. Interestingly, these various philosophies were somewhat forgotten in later historical reconstructions of the origins of the concept. Of the two that I discuss in this section, the first one was mentioned in Roberts's dissertation in 2006 and the second one in the 2005 article by Woodhouse and Breyman, but they were never discussed together in a systematic manner; they turned into mere footnotes of the history of green chemistry.[32] I argue that, for many reasons, these alternatives should not be forgotten since they offer insights lacking in the version of green chemistry that succeeded.

3.1. Hancock's environmental green chemistry

One year before the 1994 book on the benign by design philosophy, the first major American article with "green chemistry" in its title was published in March 1993 in *Science*.[33] The article presented, however, green chemistry as something slightly different from what it would become in Anastas's papers. It appears that some of the early leaders of the green chemistry movement, and in particular

---

31 Roberts, "Creating Green Chemistry," 30.
32 Edward J. Woodhouse and Steve Breyman, "Green Chemistry as Social Movement?," *Science, Technology, & Human Values* 30 (2005): 199–222.
33 Ivan Amato, "The Slow Birth of Green Chemistry," *Science* 259 (1993): 1538–1541.

Kenneth Hancock from the National Science Foundation (NSF), had a distinct vision of the field.

The article of Ivan Amato entitled "The Slow Birth of Green Chemistry" was summarized by Jody Roberts in the following way:

> The report deals explicitly with future directions of research in chemistry, and talks at length about the NSF program initiated by Kenneth Hancock that provided the seed money for the research that eventually went to press in the volume Benign by Design. And Amato discusses the links between the NSF and the EPA in establishing this program. However, Amato does not actually refer to any of this as 'green chemistry' except in the title of the article. Rather, early phrases such as environmentally benign chemical synthesis run alongside discussions of environmental chemistry and the chemical industry's Responsible Care® program.[34]

Not only was the term green chemistry barely mentioned, but the concepts of "environmental chemistry" and "environmental science" were referred to multiple times throughout the entire paper; this was the disciplinary framing of these activities. Amato quotes Mario Molin from the MIT:

> Historically, environmental science was discredited ... because people working in that field usually were not the ones making contributions to the academic community.

The article continues:

> many chemists saw it as soft, minor-league science, compared to their own hard, detailed work on chemical synthesis, catalysis, and reaction mechanisms ... environmental chemistry doesn't have to be limited to issues such as tracking contaminants, ... they're largely a reflection of yesteryear's focus on pollution abatement and remediation. To chemists, says Hancock [Kenneth Hancock, director of the National Science Foundation's chemistry division] those buzz words conjure moving dirt with heavy machinery, not synthesizing molecules or uncovering precise reaction mechanisms with elegant experimental protocols.

Amato explains that this perception began to change when environmental chemists shifted from pollution abatement to pollution prevention. "Environmental chemistry is finally becoming recognized for the sophisticated subject that it is" Amato quotes William Glaze, editor of the *Environmental Science and Technology* journal.[35]

It should be clear now that none of these scholars saw themselves as forging a new discipline and even less as outlining some paradigm shift for chemistry

---

34 Roberts, "Creating Green Chemistry," 72.
35 Amato, "The Slow Birth of Green Chemistry," 1540.

in general. Green chemistry (mentioned only in the title) appears in the article to be a renewal of a somewhat eclectic branch of environmental chemistry by making it more attractive to chemists in academia (and not to business, as in the case of Anastas's green chemistry). Amato explains that "Environmental chemistry was linked to the pollution control philosophy of the 1970s and '80s, which focused on tracking, recycling and detoxifying contaminants." Green chemistry would be then a new philosophy for environmental chemistry. The article ends by stating that environmental chemistry is still a young science, only 20 years old (in 1993), and that new funding and especially new teaching programs will make it flourish.

This way of understanding green chemistry was further formalised by Kenneth Hancock, the scientist from the NSF interviewed by Amato. He published his own programmatic and conceptual article in the same volume as Anastas in 1994. After Anastas's introductory chapter I discussed in the previous section, there was another chapter by Hancock and his colleague from the NSF, Margaret A. Cavanaugh, entitled "Environmentally Benign Chemical Synthesis and Processing for the Economy and the Environment."[36]

First of all, Hancock's and Cavanaugh's chapter is much more general in ambition than Anastas's. It starts by describing the general problems with the image of chemistry due to environmental pollution it caused; the authors discuss notably stratospheric ozone depletion, global warming, and toxic spills to the Rhine river. Hancock and Cavanaugh agree with Anastas on the issues such as the profitability of benign chemical synthesis for the industry, and the win-win logic of the new approach (the industries would save money because they would not spend it on waste treatment). The difference is that their chapter makes a much stronger moral argument in favour of sustainability.

> In addition to the many sound health and economic reasons to worry about the environment, there is another urgent moral imperative: we have a responsibility to preserve the environment and a fair share of its natural, non-renewable resources for our children and for the generations that follow.[37]

Roberts has pointed out that Anastas and Warner were much less prone to use ethics-based arguments when addressing environmental concerns (again, it

---

36 Kenneth G. Hancock and Margaret A. Cavanaugh, "Environmentally Benign Chemical Synthesis and Processing for the Economy and the Environment," in *Benign by Design: Alternative Synthetic Design for Pollution Prevention*, ed. Paul Anastas and Carol A. Farris (Washington, D.C.: ACS Symposium Series, 1994), 23–30.

37 Hancock and Cavanaugh, "Environmentally Benign Chemical Synthesis," 24.

remains unclear whether this was a way of selling the concept to the private sector, or an integral part of their philosophy).[38] Hancock's and Cavanaugh's paper is rooted in older works in environmental chemistry from the 1970s in which preservation of the natural environment was the goal in itself.

However, the differences between the two approaches do not end here. For Anastas, the benign-by-design philosophy was a way of making environmental concerns attractive to synthetic chemists. Hancock and Cavanaugh explain in their paper, however, that the NSF program on environmentally benign chemical synthesis was

> explicitly designed to encourage interdisciplinary interactions of all kinds. The idea is to stimulate the breakthroughs that come with cross fertilization by research partners of different backgrounds and perspectives. It aims to stimulate partnerships not only between industry and academe, but also between chemists and engineers, or between chemists, engineers and microbiologists, ecologists, or whoever has the needed complementary expertise.[39]

As already explained in Amato's 1993 *Science* article, for Hancock and Cavanaugh, the NSF and EPA did not try to create any new discipline, but their studies were fully in line with big themes of environmental chemistry in general:

> The research challenges to chemists in environmental chemistry are not limited to benign chemical synthesis, though synthesis and processing are clearly the base of the internationally profitable chemical industry. ... Chemists will develop the new catalysts for energy- and resource- efficient synthesis and will devise the biomimetic schemes for energy harvesting and bioremediation. Chemists will invent new disposal strategies for hazardous wastes of all types ~ through encapsulation, biodegradation, and incineration.[40]

The last ambition is particularly important. Waste disposal and treatment were still a part of the same continuum as pollution prevention.

This competing vision advanced by the National Science Foundation scientists was nevertheless brought to an abrupt end after the sudden death of Hancock in 1993 during a conference held in Budapest. A short obituary prefaced the 1994 book:

---

38 Roberts, "Creating Green Chemistry," 93.
39 Hancock and Cavanaugh, "Environmentally Benign Chemical Synthesis," 26.
40 Hancock and Cavanaugh, "Environmentally Benign Chemical Synthesis," 27.

Ken's commitment to green chemistry and pollution prevention has prompted the establishment of the Kenneth G. Hancock Pollution Prevention Fund to be administered by the American Chemical Society's Division of Environmental Chemistry, Inc.[41]

It is true that Hancock never used the term "green chemistry" in the 1994 chapter (there is, however, both "green science" and "green technology"), but he figures, nevertheless, in the collective memory as one of the key pioneers in the field, even though his green chemistry was distinct from Anastas's.

Hancock and Cavanaugh's chapter, as well as Amato's article, weaken, again, the supposed demarcation I mentioned in the previous chapter between environmental pollution chemistry (or the 'command and control' approach) and green chemistry. For the NSF scientists, the focus on benign synthesis was a promising evolution of methods that would make environmental chemistry a fully fledged respected discipline after 20 years of struggles for recognition. However, things did not turn out that way. Green and environmental chemistries are today considered two distinct fields of study. Pushed to the extreme, environmental chemistry is understood as the chemistry of the natural environment, especially of hydro- and atmosphere, while green chemistry is the chemistry of pollution prevention on the molecular level. The separation of the two is clearly visible in the popular understanding of their relationship. For example, from since January 2005, the article in English-speaking Wikipedia on "environmental chemistry" explains that the discipline "is the scientific study of the chemical and biochemical phenomena that occur in natural places [but] it should not be confused with green chemistry, which seeks to reduce potential pollution at its source."[42] Such a separation was not the case only ten years prior. For the NSF scholars in the early 1990s, green chemistry was not only a part of environmental chemistry but a hope for its renewal! And yet both terms shifted their meaning in the following years and parted away. Interestingly, there was a smooth transition between them in terms of popularity. In the English-speaking literature, green chemistry appears to have supplanted the environmental one (Figure 1).

---

41 Anastas and Farris, *Benign by Design*, xi.
42 This quotation comes from the article version as of 5 May 2021. The first instance of a Wikipedia article on environmental chemistry mentioning green chemistry in a slightly modified form comes from 23 January 2005, accessed 18/02/2022, https://en.wikipedia.org/w/index.php?title=Environmental_chemistry&diff=11018353&oldid=9584656.

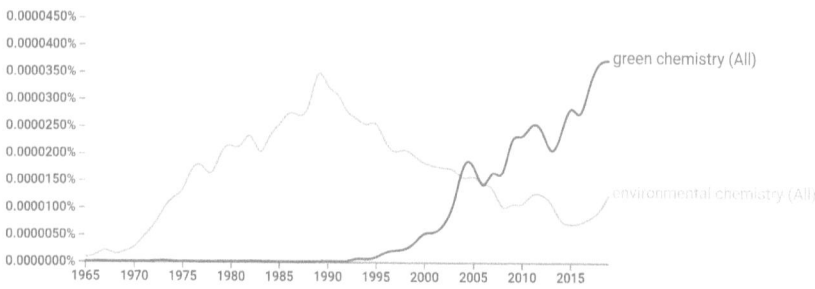

Figure 1: Evolution of the popularity of terms "environmental chemistry" and "green chemistry" in books digitized by Google according to Google Ngram (smoothing=0, English, 1965–2019)

Figure 1 shows that environmental chemistry was at the popularity peak around 1990 but was displaced by green chemistry by 2005 after a long decline. Is there a practical consequence? It is possible that young environmentally-minded chemists may be more inclined to direct themselves towards the fashionable green chemistry than towards the seemingly less exciting, and somewhat declining, chemistry of natural environment? This may mean that there would be more specialists in pollution prevention than in pollution remediation, which may not be a bad thing, under the condition that green chemistry fully delivers on its promises. A somewhat crude simplification would be to say that since green chemistry came to mean prevention and environmental chemistry came to mean the study of chemical aspects of atmosphere and hydrosphere, there is no more popular intuitive buzzword for pollution remediation (in the 1980s it was all simply environmental chemistry). The possible consequences of this shift from the sociology of knowledge point of view are yet to be understood.

Finally, all of this prompts the question of whether this strict separation was inevitable and whether Hancock, if it was not for his premature death, could not keep the benign-by-design philosophy closer to its environmental roots. We should nevertheless be cautious with this type of speculations, especially considering that there were more green chemistries fighting for recognition.

3.2. Garrett's toxicological green chemistry

A different kind of tension, and a different type of green chemistry, can be found in a 1996 book edited by Stephen C. DeVito and Roger L. Garrett under the title

Designing Safer Chemicals. Green Chemistry for Pollution Prevention.[43] Garrett was Anastas's boss when he arrived at the EPA in the early 1990s. As with the 1994 and 1996 publications edited by Anastas, this volume was also based on an earlier symposium sponsored by the ACS Division of Environmental Chemistry. Interestingly, Anastas and his works are not mentioned at all in Garrett's introductory chapter, even though its main subject is the definition of green chemistry and the discussion on its future. In fact, the chapter entitled "Pollution Prevention, Green Chemistry, and the Design of Safer Chemicals" significantly differs from what was presented in Anastas's volume from the same year. The competition between the two visions of green chemistry was noted by political scientists Edward Woodhouse and Steve Breyman in their socio-historical article from 2005:

> The green label actually encompassed at least two partially competing approaches, one of which has gained primacy to date. The road not taken was championed by Roger Garrett and Steven DeVito, midlevel scientists and administrators at the EPA, who coedited a 1996 book bearing the title Green Chemistry ... They envision the day, far from now, when chemicals are designed so as to not be assimilable into living organisms, or are designed to be easily and quickly metabolized and excreted. Medicinal chemists have long been interested in such mechanisms for pharmaceuticals, but there had been essentially no carryover from medicinal chemistry to industrial, agricultural, and other chemistries. ... Garrett predicted in 1998 that this was on the verge of changing, proposing, "Organic chemistry textbooks a generation from now will be unrecognizable compared with today's standard texts." ...
> In fact, however, green chemistry conferences and publications to date display little movement toward the difficult, sophisticated approaches outlined by Garrett and DeVito. Instead, the field has come to be dominated by a version of green chemistry closer to chemists' traditional ways of doing things, one based on alternative synthetic pathways, solvent substitution, and other tactics that attempt to circumvent the need for understanding how living organisms metabolize ingested chemicals. This second approach to green chemistry focuses on making chemicals and their production inherently benign rather than attempting brilliant metabolic maneuvers that protect living organisms from chemicals that would be dangerous if not so cleverly designed.[44]

This is a bold and, effectively, drastically different vision of what green chemistry should be. Garrett considered green chemistry in 1996 as an overarching revolutionary paradigm. His introductory chapter explains that:

---

43 Stephen C. DeVito and Roger L. Garrett, *Designing Safer Chemicals Green Chemistry for Pollution Prevention* (Washington D.C.: American Chemical Society, 1996).
44 Woodhouse and Breyman, "Green Chemistry as Social Movement?," 205.

This new approach can be considered a new paradigm that will challenge the way we have historically viewed and designed chemicals to meet the demands of the continually expanding domestic and world markets. More specifically, it will challenge the way we have dealt with environmental and human health problems caused by industrial and commercial chemicals. In order to meet the aforementioned challenges, it will necessitate changes in the current emphasis, beliefs, policies, procedures and infrastructure in both academia and industry.[45]

Interestingly, unlike Anastas and Hancock, Garrett does not make any concession to business and does not advocate a win-win situation. On the contrary, he says that

> Industry must meet the challenge by studying the concept and initiating paradigm shifts across the broad spectrum of its development and use of chemicals. The emphasis on designing safer chemicals must rank with the time-honored considerations of efficacy, cost and market share of chemical products.

Garrett notes later that the shift to this new way of thinking "undoubtedly will require substantial investments in time, money and human resources."

The chemistry of the future may be more expensive but the price is worth paying.

While Garrett's ambitious project significantly differed from both Hancock's and Anastas's, there was one common element between him and Hancock that was framed differently by Anastas. Both Garrett and Hancock sought to root their project in older scientific trajectories. For Hancock, the background discipline was environmental chemistry. For Garrett it was toxicology. According to him, is the toxicity of chemicals for humans and the environment that should be the starting point for the entire discussion on the direction that chemistry should take. On the one hand, he underlines the importance of the lifecycle assessment of chemical products (he insists somewhat more on this aspect than either Anastas or Hancock), on the other hand, he highlights the importance of education of the new generations of scientists:

> To provide adequate training of synthetic chemists interested in designing safer chemicals and destined for careers in both academia and private industry, it is believed that new curricula should be developed to provide firm grounding in biochemistry, pharmacology and toxicology. At the graduate level this may be best accomplished through joint

---

45 Roger L. Garrett, "Pollution Prevention, Green Chemistry, and the Design of Safer Chemicals," in *Designing Safer Chemicals Green Chemistry for Pollution Prevention*, ed. Stephen C. DeVito and Roger L. Garrett (Washington D.C.: American Chemical Society, 1996), 2–15, 3.

appointments and multi-disciplinary graduate committees comprised of the appropriate fields of study to oversee curricula and graduate research efforts directly related to the chemistry/biology relationships involved in designing safer chemicals.

Garrett's overarching ambition is best summarized in the following passage:

> The introduction of these new multi-disciplinary curricula and training will result in the emergence of a new "hybrid chemist" or perhaps, the "toxicological chemist." ... this hybrid chemist will evolve from the current subspecialties in synthetic chemistry. ... The new hybrid chemist or the "toxicological chemist" must consider both the function of the chemical in its industrial or commercial application and its toxicological effects in humans and the environment. In most respects achieving the delicate balance between safety and efficacy will undoubtedly prove to be the most difficult and challenging effort in the history of synthetic chemistry.[46]

Garrett also saw his green chemistry as growing from earlier debates on the place of toxicology in chemistry:

> The idea of designing safer chemicals is not new. It was the subject of a symposium held in Washington, D.C. in 1983 sponsored jointly by the EPA, Oak Ridge National Laboratory and the Society of Toxicology. In a paper presented at this symposium ... the design of safer chemicals was referred to as the "domestication of chemistry" by Dr. E.J. Ariens, the noted medicinal chemist with exceptional expertise in organic chemistry, molecular toxicology and pharmacology. The use of the term "domestication" is ideal. ... Although the idea of designing safer chemicals is not new, the concept of how this idea can be developed, introduced and integrated into the real world of commercially viable industrial chemicals is new and is the subject of this chapter and indeed this book.[47]

Everhardus Jacobus Ariëns was a Dutch pharmacologist and, looking from Garrett's perspective, the godfather of what the new green chemistry was supposed to become.

If Hancock's vision was cut short by his sudden death, Woodhouse and Breyman offer the following explanation of the victory of Anastas's vision over Garrett's:

> This direction seems to have emerged primarily because industrial chemists are trained to play around with various ways of putting molecules together—and are not trained in much having to do with humans or other living things. Moreover, it is simply easier to avoid using a dangerous chemical or chemical pathway than it is to create a sophisticated, pharmaceutical like molecular structure that will be quickly metabolized and/or excreted.

---

46 Garrett, "Pollution Prevention, Green Chemistry, and the Design of Safer Chemicals," 13.
47 Garrett, "Pollution Prevention, Green Chemistry, and the Design of Safer Chemicals," 6.

Then they turned to microsociological explanations:

> The crucial role in catalyzing green chemistry has been that of the director of the green chemistry program within the Office of Pollution Prevention and Toxics at the EPA. Had the job been held by one of those championing safer chemicals in the pharmaceutical tradition, there would have been some chance that the esoteric-chemical route would have been emphasized more equally. But Paul Anastas, the person hired to run the program at the EPA, was himself trained in standard rather than metabolic chemistry, and he chose to nurture the fledgling green chemistry enterprise along the plain-old chemistry route.[48]

Finally, it must be mentioned that Garrett passed away after a long disease in 2003, meaning that this alternative green chemistry also lost its main proponent.

3.3. Concluding remarks

Overall, Anastas, Hancock, and Garrett presented three distinct philosophies and possible lines of developments for green chemistry in the United States in the 1990s (Table 5). None of them was strictly superior, from the scientific point of view, to the others and all of them had some institutional and conceptual backing. Or to put it differently, in the middle of the 1990s, it would not be possible to tell which one of them would become the green chemistry by default. What is staggering is the scale of Anastas's victory. Very little of Garrett's toxicological approach remains in the green chemistry of today, and his ambitious vision of transforming the entire chemical enterprise is still a far cry from the way chemistry developed in general. As for Hancock's broad environmental approach, some similar ideas are being brought back through the sustainable chemistry philosophy (discussed in later chapters), but these two have a very different genealogy. As such, Hancock's legacy played little role in the development of green chemistry even though he was the pioneer of pollution prevention in chemistry.

I discussed these three green chemistries because they developed within the same group of researchers and reached a certain level of formalization through publications. However, less formalized philosophies on what green chemistry should have become were circulating as well. For instance, I can mention Joseph Breen, the first director of the Green Chemistry Institute and, like Garrett, Anastas's former boss. Breen edited an extensive book on pollution prevention in 1992, laying down some ideas on the concept even if he did not articulate its

---

48 Woodhouse and Breyman, "Green Chemistry as Social Movement?," 206.

relationship to chemistry.[49] What he heavily emphasised in this early work was the importance of ethics in industrial pollution prevention. He simply stated that "Pollution prevention is the environmental ethic of the 1990s." If so, the green chemistry would be the new ethic of the chemical industry; a dimension somewhat neglected in many further publications. It appears that his own philosophy would link green chemistry to the concept of industrial ecology.[50] Unfortunately, Breen did not have a chance to develop this line of thought in the Green Chemistry Institute due to his sudden death in 1999. Again, we are left with speculations on how would have the field developed if he had continued to shape it. In chapter 7, I explore the relationship between green chemistry and industrial ecology in more detail.

Table 5: Principal American green chemistries in the 1990s

	Anastas	**Hancock**	**Garrett**
Discipline of reference	Synthetic chemistry	Environmental chemistry	Pharmacology, toxicology
Relationship with the industry	Green chemistry must be business-friendly to be successful	Green chemistry must be business-friendly to be successful, but the moral argument is also important	The industry has to adapt to the changing paradigm even if it is costly
Key early texts	Anastas, 1994 and Anastas and Williamson, 1996	Hancock and Cavanaugh, 1994	Garrett, 1996
Relationship with environmental chemistry	Environmental chemistry is the old paradigm. Green chemistry is the new way of thinking.	Green chemistry is the natural development of environmental chemistry	Not mentioned at all.

---

49 Joseph J. Breen and Michael. J. Dellarco, *Pollution Prevention in Industrial Processes: The Role of Process Analytical Chemistry* (Washington D.C.: American Chemical Society, 1992).
50 Paul Anastas, "Joe Breen – heart and soul of Green Chemistry," *Green Chemistry* 1 (1999): G87.

Definition of green chemistry	Strong theoretical orientation, concept-heavy.	Theory-light, general directions outlined	Theory-light but clear orientation.

4. Sheldon's and Trost's green chemistry metrics

The underlying idea of this chapter is to describe different ways in which the name "green chemistry" was used in the 1990s and not to discuss similar concepts, no matter how important, that may have contributed to the emergence of the field later. There are two reasons for this approach. Firstly, I want to avoid retrofitting various earlier ideas into the later definitions, which would deform this historical reconstruction by implying that the green chemistry concept was more influential and more widespread than it really was. And secondly, the green chemistry umbrella became so broad by the beginning of the 2020s, that the study of all possible concepts that are today considered green would go largely beyond the scope of this book.

In this short section, I make an exception to the philosophy presented above as it is necessary to mention two scientists without whom the later history of green chemistry would be impossible to understand: Barry Trost and Roger Sheldon. Indeed, they did not make part of the early EPA/NSF green chemistry initiatives and they did not describe their work as green chemistry in the early 1990s. I deviate from the principle, however, because their works were not retroactively qualified as green chemistry due to some theoretical developments, but because they both enthusiastically embraced the label and became prominent members and promoters of the green chemistry movement. Their early works are an essential part of the 'green chemistry canon' since the mid-1990s.

Barry Trost, professor of chemistry at Stanford, published in 1991 in *Science* an article entitled "The Atom Economy – A Search for Synthetic Efficiency."[51] Anastas and Warner used the term Atom Economy to describe their principle 2: "Synthetic methods should try to maximize the incorporation of all materials used in the process into the final product. This means that less waste will be generated as a result." In fact, Anastas referred to Trost already in his 1994 chapter.[52] Because Trost's article came out in 1991, one scholar thought that the celebration

---

51 Barry Trost, "The Atom Economy – A Search for Synthetic Efficiency," *Science* 254 (1991): 1471–1477.
52 Anastas, "Benign by Design Chemistry," 10.

of 25 years of green chemistry in 2016 referred specifically to this publication.[53] The mistake is fully understandable in this case though. Popular from the beginning, Trost's concept grew more and more significant among scholars identifying with green chemistry throughout the 2000s when problems concerning the evaluation of what counts as green became apparent. If Anastas is considered to be the father of green chemistry, Trost is the father of green chemistry metrics.

One of the first manuals in green chemistry metrics describes Trost's contribution as follows: "Trost articulated the concept of atom economy, which characterizes the 'greenness' of a synthetic process by calculating the number of atoms from all of the reactants that make it into the final product."[54] In an ideal reaction, fully satisfying the atom economy, all atoms are used for the synthesis of the desired product. No 'waste' atoms are left. Moreover, Trost briefly explains in the original article from 1991 that: "Major benefits that derive from such processes include more effective use of limited raw materials and decreased emissions and waste disposal."[55] In other words, the 'green' ambition was explicit in Trost's work from the beginning.

The atom economy can be calculated using the following equation:

AE = X/Y * 100%, where AE stands for Atom Economy, X for the mass of atoms in desired products, and Y for the mass of atoms in reactants.

It is a quick, easy, and intuitive measure of whether a given reaction is efficient or not. It is important to mention, however, that while this calculation became ubiquitous in green chemistry metrics books and articles in the 2000s, it was not a part of the original 1991 article. In his article, all reactions presented by Trost had simply 100% Atom Economy. In this sense, the original atom economy was an ideal, not a metric properly speaking.

Green chemistry metrics has another father as well: Roger Sheldon. Sheldon, for many years professor at the University of Delft, actively published in the pages of the *Green Chemistry* journal after 1999 and presented many original and interesting ideas about what green chemistry is and should be. He had been involved in the studies and activities in line with what the EPA later defined as

---

53 John Andraos, "Useful Tools for the Next Quarter Century of Green Chemistry Practice: A Dictionary of Terms and a Data Set of Parameters for High Value Industrial Commodity Chemicals," *ACS Sustainable Chemistry and Engineering* 6 (2018): 3206–3214.
54 Alexei Lapkin and David J. C. Constable, *Green Chemistry Metrics: Measuring and Monitoring Sustainable Processes* (Wiley-Blackwell, 2008), 29.
55 Trost, "The Atom Economy – A Search for Synthetic Efficiency," 1471.

green chemistry already in the 1980s and 1990s. Sheldon even presents his own vision of these early days in one 2014 article:

> The term green chemistry emerged in the early 1990s but this doesn't mean that green chemistry had not existed before then, merely that it didn't have that name. Indeed, the drive towards waste minimisation in fine chemicals manufacture was already well underway in the 1980s, when it was widely referred to as clean chemistry, but it gathered momentum in the early 1990s with the advent of the term green chemistry and the underlying concepts of atom economy and E factors (kgs waste/kg product).[56]

There are two interesting things about the citation above. Firstly, Sheldon insists on the continuity of ideas between the 1980s and the 1990s. Green chemistry is not presented as a game-changer, but more of a new name for the tendencies well under development already before. This confirms my earlier findings. Secondly, he claims that atom economy and his own E-factor are "underlying concepts" of green chemistry, which is a much more controversial statement.

Sheldon explains that

> The E factor, … is the actual amount of waste produced in the process, defined as everything but the desired product. It takes the chemical yield into account and includes reagents, solvents losses, all process aids and, in principle, even fuel … A higher E factor means more waste and, consequently, greater negative environmental impact. The ideal E factor is zero. Put quite simply, it is kilograms (of raw materials) in, minus kilograms of desired product, divided by kilograms of product out. It can be easily calculated from a knowledge of the number of tons of raw materials purchased and the number of tons of product sold, for a particular product or a production site or even a whole company.[57]

The relation between Atom Economy and E-factor was described in the following way by green chemistry metrics experts:

> It is useful to note that the E factor mathematically incorporates all the information provided by atom economy into a larger, more encompassing mass based metric … Some of this information is not usually reflected by the atom economy metric since the E factor typically measures large solvent masses (destined for waste). Although atom economy considerations may motivate research into more atom-efficient reactions, the E factor advances a global approach to waste reduction at all stages of experimental design.[58]

---

56 Roger Sheldon, "Green and Sustainable Manufacture of Chemicals from Biomass: State of the Art," *Green Chemistry* 3 (2014): 950–963, 950.
57 Roger Sheldon's personal website, accessed 19/02/2022, https://www.sheldon.nl/roger/efactor.html.
58 Andrew P. Dicks and Andrei Hent, *Green Chemistry Metrics. A Guide to Determining and Evaluating Process Greenness* (New York: Springer, 2014), 5.

Calling the E-factor and atom economy the underlying concepts of green chemistry remains, nevertheless, problematic. It is true that both of them were the default green chemistry metrics from the outset of the field. However, what they measure is merely the amount of waste. They remain silent on their toxicity, they do not evaluate the safety of handling chemical substances in question, they do not incorporate the problem of renewability, and, broadly speaking, they do not concern sustainability. They are completely irrelevant for more than half of the 12 principles. Why did they become so popular then? The most likely hypothesis is directly related to the supremacy of Anastas's green chemistry. In the 1996 book, Anastas and Williamson were fully aware of the atom economy's shortcomings and its narrow scope of application, but they glossed over the issue and still called it "one of the central tenants [sic] for any synthesis that is striving toward Green Chemistry."[59] It is possible that for Garrett and his ecotoxicological chemistry, the default metrics would be found in toxicological studies, and for Hancock, with his broad approach environmental approach, some form of the life cycle assessment would probably take precedence. But Anastas's chemistry was rooted in synthetic chemistry. Atom economy, in particular, is the prime example of an elegant solution with which a synthetic chemist can shine and show his or her skills. The clever use of carefully selected reactants and catalysts makes it possible to drastically reduce the overall waste output. This is also a skill easily marketable to the industry. Instead of lengthy, expensive, and inevitably ambiguous life cycle assessments, or even more politically charged discussions on what a sustainable process is, the atom economy and E factor concepts are easy to explain. They permit a 'single player' (neoliberal?) thinking about the problem of wastes and pollution instead of a systemic one. And while the 12 principles developed by Anastas and Warner go far beyond this narrow understanding, the core elements of the EPA's green chemistry as developed by Anastas are rooted in synthetic chemistry and business friendliness. If it was not Anastas's green chemistry that came to dominate the field, neither Trost's nor Sheldon's metrics might have ever gained such popularity.

However, there is one more layer to all of this. In 1992, Sheldon published a chapter entitled "Catalysis, the Atom Utilization Concept and Waste Minimization" in a collective work *Industrial Environmental Chemistry: Waste Minimization in Industrial Processes and Remediation of Hazardous Waste* issued from

---

59 Anastas and Williamson, "Green Chemistry: An Overview," 14.

a symposium organized in March that year.[60] The volume's editors, Donald T. Sawyer and Arthur E. Martell, do not dwell on theoretical debates in its introduction but briefly explain the ambition of the book in the following way:

> The subject of this conference reflects the interest that has developed in academic institutions and industry for technological solutions to environmental contamination by industrial wastes. Progress is most likely with strategies that minimize waste production from industrial processes. Clearly the key to the protection and preservation of the environment will be through R&D that optimizes chemical processes to minimize or eliminate waste streams.[61]

In other words, this "industrial environmental chemistry" was very close in its ambitions to what the EPA's benign-by-design chemistry aspired to become over the same period.

Sheldon talks in his chapter about the atom utilization concept that largely parallels Trost's atom economy. It seems that Sheldon developed an almost identical idea at the same time independently. But Sheldon talks also about waste prevention involving new synthetic pathways and he underlines the role of catalysis in this process. These are some crucial components of what would become the 12 principles of green chemistry. In this sense, Sheldon's ideas are green chemistry *avant la lettre* in a very direct sense. If the industrial environmental chemistry community had fully embraced Sheldon's ideas and transcribed them into a more formalized project, he would have been considered the true father of what is today labelled as green chemistry. This is perhaps why Sheldon, even though fully embraced the term green chemistry and defended its use on multiple occasions because of how it resonates with the wider public,[62] showed some more hesitation during the 25th anniversary when it comes to its origins. He was one of very few scientists who mentioned Pavel Drašar's 1991 article on green chemistry to illustrate the fact that the time was ripe for such a vocabulary to emerge all over the world, and that the EPA did not have a monopoly over it.[63]

---

60 Donald T. Sawyer and Arthur E. Martell, ed., *Industrial Environmental Chemistry: Waste Minimization in Industrial Processes and Remediation of Hazardous Waste* (New York: Springer, 1992).
61 Sawyer and Martell, *Industrial Environmental Chemistry*, vii.
62 Roger Sheldon, "Editorial," *Green Chemistry* 2 (2000): G1–G2.
63 Roger Sheldon, "Green chemistry and resource efficiency: towards a green economy," *Green Chemistry* 18 (2016): 3180–3183.

5. Clark's Green Chemistry

The late 1990s were marked by four catalytic events in the history of mainstream green chemistry as we know it today. Firstly, in 1997, the Green Chemistry Institute, a non-profit organization of which the first executive director was Joseph Breen, was established. The Institute was a collaborative initiative involving entities such as the EPA, the University of North Carolina, and companies such as Hughes Environmental and Praxis. In 2001, the Institute became a part of the American Chemical Society. In 2004, Anastas became its full-time director after leaving a position at the Office of Science and Technology Policy in President Bush's White House.

The second catalytic event also took place in 1997: the first international conference in green chemistry in Venice sponsored by the International Union of Pure and Applied Chemistry (IUPAC), whose working group on green chemistry had been created one year prior, and co-sponsored by the EPA and the American Chemical Society. Anastas and Breen attended the conference as keynote speakers.[64]

The third catalytic event was the already mentioned publication of the seminal book *Green Chemistry: Theory and Practice* by Anastas and Warner in 1998. The book presented the 12 principles of green chemistry which codified the previous reflection developed in the EPA (in line with Anastas's philosophy) as well as expanded it with new elements. This foundational textbook is still a reference in the field.

The final catalytic event took place in 1999: the *Green Chemistry* journal was established. *Green Chemistry* remains until today the most important and influential review that sets the tone for all debates on the topic of green chemistry.

There is then an institution, an international conference, a textbook, and a journal; four elements necessary for a scientific discipline, properly speaking, to emerge. All four were established in the same short time frame giving the impression that they all belong to the same trajectory. And yet, one of these is not like the others. In fact, the journal was not established by EPA scholars but by a British chemist from the University of York, James H. Clark. Clark was not part of the EPA's network. Besides, he was barely familiar with the EPA's green chemistry framework. Linthorst (2010) explains Clark's situation in the late 1990s in the following way:

---

64 Mervyn Richardson, "International Conference on Green Chemistry: Challenging Perspectives," Chemistry International 19 (1997): 187–189.

he published an article on the catalysis of the Knoevenagel reaction in Reactive & Functional Polymers … In this article he used the term green chemistry one time (in the keywords), but he did not explain this term. In fact, he didn't propose, or referred to, a particular green chemistry philosophy. But his work was intensively cited, which was not the case for the US EPA works of that time. Clark continued to use the term in the same way. In 1998 he wrote an article in Chemical Communications …, which was cited approximately 210 times! Considering ISI Web of Knowledge: Web of Science, until 1998 Clark never referred to the green chemistry philosophy developed within US EPA. Clark also focused on networking … This resulted in the founding of the Green Chemistry Network (GCN) within the Royal Society of Chemistry (RSC). The proposed activities of GCN were located within the University of York and Clark became the first director of this network. Alongside the establishment of GCN, RSC announced the launching of the new journal Green Chemistry with Clark as its first editor.[65]

Why did the Royal Society of Chemistry choose the name *Green Chemistry* for its new journal? Linthorst continues:

> Considering the choice of the title of the journal, Clark argued that the term green chemistry carried a good "combination of widespread use and appreciation, as well as simplicity and impact" … Terms like "widespread use," "simplicity" and "impact" outline that the type of term, or say language, was important in choosing the title of the new journal, but what about the role of knowledge. In the first Editorial the editor referred to the handbook of green chemistry developed within US EPA. The (first) scientific editor of Green Chemistry also wrote the first article on green chemistry in this journal …, which was based on his Inaugural Lecture at the University of York in 1998. However, the editor did not refer to the previously proposed green chemistry works of US EPA at all.[66]

It is true that in the first article of the first issue of *Green Chemistry*, Clark uses the term numerous times but he never defines it.[67] For example, the article's summary states that "The green chemistry revolution is providing an enormous number of challenges to those who practice chemistry in industry, education and research" and that in this article Clark "considers some of the new and successful "greener" chemistry in practice." Some sort of limited definition is in Clark's short biography in the same article: "His research interests cover various aspects of Green Chemistry including clean synthesis and new materials," but this is not very specific. While he mentions the rise of green chemistry

---

65 Linthorst: "An overview: origins and development of green chemistry," 61.
66 Linthorst: "An overview: origins and development of green chemistry," 63.
67 James Clark, "Green chemistry: challenges and opportunities," *Green Chemistry* 1 (1999): 1–8.

conferences, networks, and awards, such as the Green Chemistry Challenge established by President Clinton in 1995 (which was co-sponsored by the EPA), it appears that his definition of what green chemistry is, and what it should be, is very intuitive and not grounded in any previous conceptual framework. There is no talk about life-cycle assessment, sustainability, or protocols for establishing or measuring greenness (no atom economy or E-factor). There is no mention of the 12 principles, in spite of the fact that the journal was established one year after Anastas and Warner's book.

Clark's green chemistry relied on a 'gut feeling' that a given process is wasteful or dangerous, and that it can be improved or 'greened.' Clark and the British Green Chemistry Network jumped on the bandwagon of green chemistry because it became a fashionable term corresponding to the *Zeitgeist* of the late 1990s. The connection to the EPA and Anastas was mostly an afterthought due to the fact the American definition of green chemistry became increasingly popular from the early 2000s on (even if Clark participated in the Gordon Research Conferences back in 1997, so some general ideas certainly circulated before).

This is not to criticize Clark or other British scholars. On the contrary, their ambitions and projects were laudable and Clark's role in greening chemistry and providing a platform for the discussion on the environmental transformation cannot be overstated. At no point in this book do I criticize the scientific and technological contributions of our story's protagonists. What I criticize is the story these protagonists tell themselves and that has been then repeated by subsequent generations of scholars without nuance. Let us recall that in 2016 the *Green Chemistry* journal celebrated the 25th anniversary of the discipline of green chemistry with monthly articles on every single of the 12 principles. With this in mind, it is important to remember that: 1) neither Clark nor any of the early contributors to *Green Chemistry* referred to the 12 principles (supposing they even knew them), 2) for none of these people the year 1991 was important in any way and they certainly did not see it as any sort of a new beginning, 3) none of these scholars believed that the purpose of the journal was to provide a platform for some proto-disciplinary body of knowledge developed in the EPA over the previous years. Green chemistry was just it: a nice buzzword for describing more environmentally friendly chemical syntheses. The fact that the *Green Chemistry* journal celebrated in 2016 the anniversary of green chemistry, an idea supposedly born in 1991 thanks to Anastas, was a result of a self-propelling narrative that dominated the field between 1999 and 2016, and which itself resulted from the constant need of retelling the history of green chemistry to draw and redraw its frontiers.

6. But what was practised as green chemistry in the 1990s?

In the sections above I mostly focused on big concepts and theoretical debates on how to define green chemistry emanating from scientists and institutions explicitly trying to address the issue. However, over the same period in the United States, more and more research projects and teaching curricula were using the name green chemistry to describe a certain type of new chemistry. In some cases, this 'every-day' use of the term developed inside the broad EPA network and sometimes completely independently due to the simple fact the name green chemistry was growing in popularity. In this section, I want to show a few examples of what was actually hiding behind the name and what kind of science was labelled as green chemistry in the US in the 1990s.

6.1. Presidential Green Chemistry Challenge Award

Perhaps the most famous practical translation of the EPA's green chemistry philosophy into the realm of research was the Presidential Green Chemistry Challenge Award launched by President Clinton in 1995 with the EPA's benediction. The first awards were announced in 1996. The award constitutes an important link between American and British green chemistries as both Anastas and Clark, along with many other scholars, considered its establishment to be the formative moment for the discipline. Clark cited the award in the first editorial of *Green Chemistry* as a model example of the type of research he wanted to explore in his journal. It was also for the first time that the term green chemistry reached a wider audience. Arguably, it was the Clinton administration's decision that propelled the development of the concept and encouraged the EPA scholars to use the name more in public communications. The award has been bestowed until this day (2022) in five categories: Greener Synthetic Pathways, Greener Reaction Conditions, Designing Greener Chemicals, Small Business, and Academic. Table 6 provides an overview of the prizes in 1996.[68]

---

[68] The award's official website, accessed 19/02/2022, https://www.epa.gov/greenchemistry/green-chemistry-challenge-winners#1996.

But what was practised as green chemistry in the 1990s?

Table 6: Presidential Green Chemistry Award recipients in 1996

Category	Recipient	Contribution
Greener Synthetic Pathways	Monsanto Company	Catalytic Dehydrogenation of Diethanolamine
Greener Reaction Conditions	The Dow Chemical Company	100 Percent Carbon Dioxide as a Blowing Agent for the Polystyrene Foam Sheet Packaging Market
Designing Greener Chemicals	Rohm & Haas	Designing an Environmentally Safe Marine Antifoulant
Small Business	Donlar Corporation	Production and Use of Thermal Polyaspartic Acid
Academic	Professor Mark Holtzapple	Conversion of Waste Biomass to Animal Feed, Chemicals, and Fuels

The problem with the award was that the selection of laureates was not based on any formalized greenness criteria in that period, and especially not on the 12 principles that were developed only later. In fact, its attribution was mostly conditioned by a very intuitive understanding of what felt green at any given moment. The result is that if we analyse the awarded research projects retrofitting them in the modern conceptualisations of green chemistry or evaluating them through various tools developed later, the supposed greenness of some of them may become questionable. Of course, it can be argued greenness should be evaluated exclusively in a given historical moment, but, nevertheless, it is crucial to at least remember this contingency.

For instance, Clark cherished back in 1999 a prize bestowed upon Monsanto for the catalytic dehydrogenation of diethanolamine. What is it?

> DSIDA is a key building block for the herbicide RoundUp®. Monsanto's novel synthesis of DSIDA eliminates most of the manufacturing hazards associated with the previous synthesis; it uses no ammonia, cyanide, or formaldehyde. This synthesis is safer to operate, has a higher overall yield, and has fewer process steps

explains the prize's website and adds that Monsanto's Roundup® is "an environmentally friendly, nonselective herbicide."[69] To be sure, I do not want to enter the debate on RoundUp's supposed ecotoxicological or carcinogenic effects.

---

69 The award's official website, accessed 19/02/2022, https://www.epa.gov/greenchemistry/presidential-green-chemistry-challenge-1996-greener-synthetic-pathways-award.

However, the point is that, as mentioned already in 2012 by the French economist Estelle Garnier, RoundUp is not what the wider public would intuitively call green precisely because of the controversies that came to surround it.[70] Not to mention the fact that some would not consider using chemical herbicides green at all.

The RoundUp case certainly indicates that there is a need for solid criteria of greenness in order to justify the attribution of the green label to this or that process or product. This need makes theoretical discussions on the concept of green chemistry particularly relevant, but these discussions often struggle to give us definitive answers. For example, it may be argued that the object of the prize mentioned above was not the herbicide but the process of its production. The new process made its manufacture more environmentally friendly and that is all that matters. However, this opens a caveat. Can the greening of a single technology contribute to the maintenance of a technical system or an industrial activity that is not sustainable and green as a whole? Can, for example, a mineral extraction technology that generates less toxic by-products be green, if the mineral in question is not renewable? One of the contributors to the first issues of *Green Chemistry* thought so, and argued that the "elimination of toxic chemicals should be a goal of green chemistry in the mining industry" even if the entire activity is not sustainable in the long run.[71] If a hypothetical fertilizer is shown to be potentially harmful to biodiversity, is reducing the amount of toxic waste generated during its manufacture a step towards greenness? This is one of the problems that has been haunting green chemistry since its first days and I discuss it more in chapter 4.

The final thing that has to be mentioned in the context of the presidential awards concerns the companies that were awarded in these early years such as Monsanto (first prize: 1996), Dow Chemical (1996), and DuPont (2003). Many former EPA administrators became later in their careers environmental directors in the companies in question.[72] This is not to suggest collusion, but, as noted Garnier, the EPA actively sought legitimacy for green chemistry among

---

70 Estelle Garnier, "Une approche socio-économique de l'orientation des projets de recherche en chimie doublement verte" (PhD diss., University Reims Champagne-Ardenne, 2012), 91.
71 Ian D. Brindle, "Green chemistry—a Canadian perspective," *Green Chemistry* 1 (1999): 155–157.
72 Andrew Hoffman, *From Heresy to Dogma: An Institutional History of Corporate Environmentalism* (San Francisco: New Lexington Press, 1997), 125.

the companies it had close ties to.[73] From the very inception, the Presidential Green Chemistry Challenge Award was deeply rooted in the American corporate culture.

6.2. First university courses in green chemistry

I previously explained that institutions, conferences, manuals, and journals are what make a discipline. And yet, there had been green chemistry university classes even before any formalization of the discipline took place. The first mention of a green chemistry course comes from a 1995 article in the *Journal of Chemical Education* by Terrence J. Collins from Carnegie Mellon University (CMU). The title of the contribution is "Introducing Green Chemistry in Teaching and Research." Collins makes there a few interesting observations on the origins of the term and the reasons for its popularity:

> In the midst of a more vigorous concern for the environment, several terms have been put forward to capture an important idea. The terms are "green chemistry," "primary prevention," and "environmentally benign chemistry." They represent the supposition that chemical processes that carry environmental negatives can be replaced with less polluting or nonpolluting alternatives; the hypothesis has been demonstrated to be true in several important cases. This idea is ethically and politically powerful. The term, "green chemistry" is now the most widely used. It is perhaps the strongest of the three because it associates developments in chemistry with the pastoral longing of modern man. The idea of green chemistry has an energy that properly belongs in university research laboratories and classrooms.[74]

"Pastoral," "ethical," and "politically powerful" are some of the terms associated with green chemistry that show the mindset of these early forerunners. Noteworthy, for Collins green chemistry appears to be an openly political endeavour, which is not necessarily what the EPA scholars had in mind.

From the article we learn that the course "Introduction to Green Chemistry" has been taught at the CMU since… 1992. It has to be reminded that no major publications using the term appeared until 1993. In this sense, the CMU course was another full-fledged green chemistry philosophy on its own merits developed independently and not related to the theoretical work within the walls of the EPA. Collins enumerates the curriculum's objectives:

---

73 Garnier, "Une approche socio-économique," 60.
74 Terrence J. Collins, "Introducing Green Chemistry in Teaching and Research," *Journal of Chemical Education* 72 (1995): 965–966.

1. To define goals for green research and to consider evaluation criteria for green reagents.
2. To identify target technologies that realistically might be replaced by green technologies and to ponder the alternatives.
3. To describe the features of polluting chemical technologies that have led to their continued use after negative environmental impacts have been discovered.
4. To discuss how successful green chemicals probably will exhibit mechanistic features similar to the reagents they are developed to replace.
5. To challenge students to think originally about what green chemistry might become.
6. To introduce students to one or more internationally recognized scholars on the chemistry or biochemistry of environmental protection.[75]

Then he proceeds to explain the actual content of the classes (Table 7).

Table 7: Topics in green chemistry at the Carnegie Mellon University, by T. Collins (1992–1995)

Topic	Number of Lectures	Description
Current examples of the role of catalysis in green chemistry	7	Catalysis in the economy; examples of primary and secondary prevention
An overview of energy sources and associated pollution	8	Technical description of fossil fuel, nuclear and solar components of the economy, and environmental impacts. Introduction to combustion and strategies for greening of combustion. Approaches to green energy sources.
Anthropogenic atmospheric pollution	5	Atmospheric chemistry of the CFC's. Nitrogen Oxides (NOx) in the Atmosphere
Biocatalysis	2	Route to new chemicals, including perhaps, commodity chemicals
Bioremediation	3	-

There are here three big lines of investigation: catalysis (to prevent pollution), alternative energies (although renewability is not mentioned explicitly), and the treatment of pollution that is already in the environment (a traditional

---

[75] Collins, "Introducing Green Chemistry," 965.

topic in environmental chemistry). Some notable features are missing: in particular concepts such as atom economy or safer solvents that were part of the EPA's green chemistry canon. On the other hand, Collin's green chemistry is explicitly dealing with pollution remediation and is rooted in the environmental chemistry conceptual framework close to the one developed by Hancock. In the bibliography cited by the end of the article, Collins mentions textbooks on environmental chemistry, environmental toxicology, and pollution. There is no trace of the EPA's green chemistry though.

Another pioneer of green chemistry education was Albert Matlack from the University of Delaware. In the first issue of *Green Chemistry*, he writes there that he felt inspired to establish a green chemistry curriculum, among others, after having attended an ACS Green Chemistry Symposium in Washington in 1994 (the one that led to the publication of the 1996 book on green chemistry co-edited by Anastas). Matlack was teaching green chemistry yearly between 1995 and 1998. How did he define green chemistry?

> Green chemistry is the chemistry of a sustainable future. A sustainable future is one that allows future generations as many options as we have today. The industrial ecology being studied by engineers and green chemistry are both parts of one approach to a sustainable future.
> On the other hand, environmental chemistry, as taught today, is largely the study of what man has put into the environment and its effect, as well as how to remediate contaminated sites.
> Green chemistry is interdisciplinary. When I lecture on waste minimisation, ion exchange resins and zeolites, I sound like a chemical engineer. When I talk about population and the environment, I sound like a physician. When I go into renewable energy, I act like a physicist.[76]

Matlack mentions sustainability and industrial ecology, two large concepts gaining popularity among environmentally minded scholars over the period. However, for Matlack, the frontiers between these various ideas are rather porous and certainly not at the heart of the debate. He describes green chemistry as an interdisciplinary endeavour, going largely beyond traditional divisions. He presents in the article the following outline of the class:

1. The need for Green Chemistry; 2. Doing without toxic chemicals (illustrated by phosgene); 3. The chlorine controversy; 4. Toxic heavy metal ions; 5. Solid catalysts and reagents for ease of work-up; 6. Solid acids and bases; 7. Separations; 8. Working without organic solvents; 9. Biocatalysis and biodiversity; 10.

---

76 Albert Matlack, "Teaching green chemistry," *Green Chemistry* 1 (1999): 19–20.

Stereochemistry; 11. Agrochemicals; 12. Materials for a sustainable economy; 13. Chemistry of longer wear; 14. Chemistry of recycling; 15. Energy and the environment; 16. Population and the environment; 17 Environmental economics; 18. Greening.

This is a rich, eclectic, and diverse curriculum, dealing with problems that 'feel' green and environmentally relevant, and that go largely beyond later articles and textbooks in which green chemistry was mostly presented through the lenses of green synthesis. In particular, topics on energy, environmental economics, and (over)population stand out. However, Matlack was not fully satisfied with the class. He wanted to go a step further. "My goal is to stop teaching green chemistry as a separate course and instead insert portions of it into all the other chemistry and chemical engineering courses." He pointed out, however, that he might have trouble convincing his colleagues to incorporate his ideas into their chemistry classes.

Matlack wrote in 2001 one of the first full-fledged green chemistry textbooks based on the structure of his class, allowing us to delve deeper into his understanding of the field.[77] Unlike Anastas and Warner whose 1998 book was above all an attempt to lay out a conceptual framework for green chemistry, Matlack's work was a much more voluminous publication structured around concrete scientific problems. It is important to underline that the book does not mention the 12 principles of green chemistry and the name of Anastas is mentioned only a few times in the references (and never in the body of the text). In other words, in the early 2000s, it was perfectly possible to still write about green chemistry without reference to the EPA's framework.

The case of the two teachers, Collins and Matlack, proves that green chemistry was not a single body of knowledge in the 1990s, but rather a set of loosely connected intuitions about what should be considered environmentally relevant, deeply rooted in the aspirations of the old environmental chemistry. In a way, just as much as there were Hancock's and Garrett's green chemistries, there were Collins's and Matlack's green chemistries. In Collins's case, it appears that his ideas developed in parallel to the ones in the EPA. Matlack, on the other hand, was directly inspired by the EPA's conferences but did not necessarily follow their more conceptual and theoretical framing.

---

[77] Albert Matlack, *Introduction to Green Chemistry* (New York: Dekker, 2001).

6.3. Canonical green chemistry symposia (1994, 1996)

In the previous sections, I talked about three books stemming from green chemistry symposia in 1993 and 1994, two of them co-edited by Anastas and one by Garrett. I focused so far on the theoretical introductory chapters that tried to define the benign-by-design and green chemistry philosophies. But the books were, of course, much richer and included a range of papers from contributors who may, or may have not, subscribed to the exact definitions presented in the introductory chapters. Table 8 shows the titles of the chapters coming from these three books.

Table 8: Overview of chapters published in canonical early green chemistry books

Paul Anastas and Carol A. Farris, Benign by Design: Alternative Synthetic Design for Pollution Prevention (Washington, D.C.: American Chemical Society, 1994).
1. Benign by Design Chemistry
2. Microbial Biocatalysis: Synthesis of Adipic Acid from D-Glucose
3. Mechanistic Study of a Catalytic Process for Carbonylation of Nitroaromatic Compounds: Developing Alternatives for Use of Phosgene
4. Preparative Reactions Using Visible Light: High Yields from Pseudoelectrochemical Transformation; A Photochemical Alternative to the Friedel—Crafts Reaction
5. Mn (III)-Mediated Electrochemical Oxidative Free-Radical Cyclizations
6. Supercritical Carbon Dioxide as a Medium for Conducting Free-Radical Reactions
7. The University of California—Los Angeles Styrene Process
8. Generation of Urethanes and Isocyanates from Amines and Carbon Dioxide
9. Nucleophilic Aromatic Substitution for Hydrogen: New Halide-Free Routes for Production of Aromatic Amines
10. Chemistry and Catalysis: Keys to Environmentally Safer Processes Alternative Syntheses and Other Source Reduction Opportunities for Premanufacture Notification Substances at the U.S. Environmental Protection Agency
11. Computer-Assisted Alternative Synthetic Design for Pollution Prevention at the U.S. Environmental Protection Agency

Paul Anastas and Tracy C. Williamson, Green Chemistry. *Designing Chemistry for the Environment* (Washington D.C.: American Chemical Society, 1996).

1. Green Chemistry: An Overview
2. New Process for Producing Polycarbonate Without Phosgene and Methylene Chloride
3. Caprolactam via Ammoximation
4. Generation of Organic Isocyanates from Amines, Carbon Dioxide, and Electrophilic Dehydrating Agents Use of o-Suifobenzoic Acid Anhydride (phosgene)
5. Clean Oxidation Technologies: New Prospects in the Epoxidation of Olefins;
6. Dimethylcarbonate and Its Production Technology
7. Selective Mono-Methylation of Arylacetonitriles and Methyl Arylacetates by Dimethylcarbonate: A Process without Production of Waste
8. Oxidation of Phenolic Phenylpropenoids with Dioxygen Using Bis(salicylideneimino) ethylenecobalt(II)
9. Kinetics of Zeolitic Solid Acid-Catalyzed Alkylation of Isobutane with 2-Butene
10. The Role of Catalysts in Environmentally Benign Synthesis of Chemicals
11. Supercritical Carbon Dioxide as a Substitute Solvent for Chemical Synthesis and Catalysis
12. Reduction of Volatile Organic Compound Emissions During Spray Painting A New Process Using Supercritical Carbon Dioxide to Replace Traditional Paint Solvents
13. Chemically Benign Synthesis at Organic—Water Interface
14. Environmentally Efficient Management of Aromatic Compounds
15. Environmentally Benign Production of Commodity Chemicals Through Biotechnology: Recent Progress and Future Potential
16. Teaching Alternative Syntheses: The SYNGEN Program
17. Incorporating Environmental Issues into the Inorganic Curriculum

Stephen C. DeVito and Roger L. Garrett, *Designing Safer Chemicals Green Chemistry for Pollution Prevention* (Washington D.C.: American Chemical Society, 1996).

1. Pollution Prevention, Green Chemistry, and the Design of Safer Chemicals
2. General Principles for the Design of Safer Chemicals: Toxicological Considerations for Chemists
3. Cancer Risk Reduction Through Mechanism-Based Molecular Design of Chemicals
4. Isosteric Replacement of Carbon with Silicon in the Design of Safer Chemicals
5. Design of Biologically Safer Chemicals Based on Retrometabolic Concepts
6. Predicting Rates of Cytochrome-P450-Mediated Bioactivation and Its Application to the Design of Safer Chemicals
7. Use of Computers in Toxicology and Chemical Design;
8. Designing Biodegradable Chemicals
9. Designing Aquatically Safer Chemicals
10. Designing Safer Nitriles
11. Designing an Environmentally Safe Marine Antifoulant
12. Imine—Isocyanate Chemistry: New Technology for Environmentally Friendly, High-Solids Coatings

A quick glance at the titles of these chapters shows immediately the difference between the two philosophies presented in the volumes edited by Anastas and his colleagues, on the one hand, and Garrett and DeVito on the other. The second 1996 conference stands apart because it includes numerous papers on ecotoxicological and health considerations, whereas the former are clearly oriented towards chemical synthesis. The papers in the second 1996 conference are also often more all-embracing and slightly more theoretical. They do not describe exclusively specific reactions but focus on the greenness of a technological approach in a given environment.

Upon deeper reading of the chapters in question, it can be observed that the vast majority of contributors to Garrett's and DeVito's volume came from academia, and in particular from the departments of pharmacology, medicine, and similar. Only two out of twelve contributors came from the industry. This is different from the chapters in the books edited by Anastas in which the participation of the industry was more pronounced with three out of eleven articles coming from authors in the private sector in 1994 and six out of seventeen in 1996. It is also worthwhile mentioning that the 1996 book edited by Anastas includes 6 chapters coming from authors located abroad (not in the US), one from Japan and five from Italy.

As for the exact topics treated in the books edited by Anastas, the vast majority of the chapters concern, unsurprisingly, alternative pathways to different chemical compounds that generate less toxic waste. Some chapters include topics that would settle at the core of the green chemistry research strategy for years, such as alternative solvents (e.g. supercritical $CO2$), but, remarkably, a huge proportion of them mention catalysis, which is presented as an obvious way for greening a great many chemical reactions. One article is a true apology for catalysis and for the role it is supposed to play in the nascent field of green chemistry.[78] The French economist Martino Nieddu and his colleagues noted a bit sceptically many years later that the emphasis early proponents of green chemistry had put on catalysis as a centrepiece of the discipline was somewhat "curious," considering that catalysis had been around for decades.[79] In defence, it

---

78 Milagros S. Simmons, "The Role of Catalysts in Environmentally Benign Synthesis of Chemicals," in *Green Chemistry. Designing Chemistry for the Environment*, ed. Paul Anastas and Tracy C. Williamson (Washington D.C.: American Chemical Society, 1996), 116–130.

79 Martino Nieddu, Franck-Dominique Vivien, Estelle Garnier, and Christophe Bliard, "Existe-t-il réellement un nouveau paradigme de la chimie verte?," *Natures Sciences Sociétés* 22 (2014): 103–113, 107.

could be argued, like Roberts did, that the originality of green chemistry lies precisely in the fact that it superposes fields of knowledge as different as toxicity studies and catalysis.[80]

Retrospectively, the articles published in the 1994 and 1996 volumes edited by Anastas satisfy some of the principles of green chemistry, notably principle 3 (less hazardous chemical syntheses), principle 5 (safer solvents and auxiliaries), and principle 9 (use of catalysis instead of stoichiometric reagents). Indirectly, they also subscribe to the general tenets of principle 1 (waste prevention), and principle 4 (designing safer chemicals). However, this is less than half of the principles. Trost's atom economy (principle 2), while lauded by Anastas in both introductory chapters, is mentioned only in one empirical article from 1994, and appears to be unknown to other contributors. Energy efficiency (principle 6) is at the margin of the discussion, not to mention renewable feedstocks (principle 7). Issues such as design for degradation (principle 10 – although this topic actually appears in Garrett's book), real-time analysis for pollution prevention (principle 11), and inherently safer chemistry for accident prevention (principle 12) are largely outside the scope of the interest of the contributors of both volumes. This is to reiterate the idea that the 1998 seminal book expanded the scope of green chemistry in the directions that had not been present in the green chemistry as practised in the 1990s even in the circles close to the EPA. Or to put it more diplomatically: there was a mismatch between the theoretical ambitions of the green chemistry project and the actual research led by people identifying with the movement.

6.4. Note on the American Chemical Society Symposia Series

Before finishing, one more unusual use of the term green chemistry should be mentioned. I explained that the Green Chemistry Institute became a member of the American Chemical Society in 2001. French economist Estelle Garnier looked in her study for the terms "green chemistry" and "green chemicals" in the ACS Symposia Series (so the series in which the previously mentioned volumes edited by Anastas and Garrett had also been published) before and after 2001.[81] Garnier found out that there was a major thematic gap between pre- and post-2001 publications in the ACS database. After 2001, the ACS publications on green chemistry used the term in a sense attributed to it by Anastas and Warner

---

80 Roberts, "Creating Green Chemistry," 9.
81 Garnier, "Une approche socio-économique," 101.

(chemical synthesis for pollution prevention and the 12 principles). However, the publications labelled with "green chemistry" keywords between 1998 and 2001 had a completely different character. Many of them concerned for example food science (e.g. food phytochemicals for cancer prevention) and agricultural chemistry in the relationship to the environment (e.g. surfactant-based separations). In the context of the latter, it was not only pollution prevention but also pollution remediation that was studied. After 2001, these topics were not associated with green chemistry any more, visibly under the influence of the Green Chemistry Institute. It is hard to position these early ACS publications and understand why they used the green chemistry label, but we can see here traces of yet another lost intuitive tradition that failed to construct its identity in a more meaningful way.

Conclusions

Green chemistry in the 1990s, especially before the popularization of the 12 principles, was marked by a great heterogeneity of approaches and research topics. Different scientists argued for divergent definitions and connected green chemistry to diverse scientific trajectories and previously developed frameworks. Teachers covered in green chemistry classes a wide range of subjectively chosen environmental topics. The term, variously interpreted, circulated in many languages bringing about different connotations.

Placing the birth of green chemistry in 1991 appears to be an almost arbitrary choice among many other possible alternatives. On the contrary, the more we dig into the past, the more shaky the foundations of this narrative are. The most interesting question stemming from this chapter is whether green chemistry could have turned out differently. Could it have been integrated with environmental chemistry and better connect to pollution remediation? Could ecotoxicology have played a more prominent role in its development? Can the failed green chemistries of yore be any help for the challenges of today? From the historian's perspective, another question is nevertheless more pressing: why did green chemistry, as conceived by Anastas, win over such a large community in the 2000s? The elegant simplicity of the 12 principles appears to offer some answers.

Chapter 3: 12 principles of green chemistry and their proliferation

1. A short epistemological introduction

The 1998 *Green Chemistry: Theory and Practice* is the most often cited green chemistry book in the history of the discipline. It is a short publication summarizing in an approachable way a philosophy that was about to gain worldwide popularity. It presents for the first time the standard historical narrative on the field with chemical catastrophes and massive pollution as the backdrop against which developed environmentalism with its command and control approach that was later put into question by the Pollution Prevention Act. The book discusses some promising research trajectories and technologies that are poised to transform the practice of the chemical industry in the upcoming years and make chemistry more environmentally friendly. The publication was thoroughly dissected and commented on by historians and chemists and it is often considered the founding text of green chemistry. However, in terms of the actual content, it is largely a continuation and reformulation of the ideas already expressed in Anastas's introductory chapters from 1994 and 1996.

Why did this book become so immensely popular as compared with quite a few other green chemistry textbooks published around the same period? One advantage was that it was short. Unlike voluminous manuals trying to outline the results of chemical research on pollution and the environment (for example Matlack's 2001 compendium[1]), the 1998 book was more of an introduction understandable for people with an elementary knowledge of chemistry. However, it seems evident that the main reason for its popularity was the fact that it formulated the so-called 12 principles of green chemistry: concise, poignant rules concerning the way chemistry should be practised in the laboratory and in the factory. I showed in chapter 1 the shortened version of the principles often reproduced in various scientific articles, in which every principle was labelled with a short name: 1. Pollution prevention; 2. Atom Economy, and so on. These names are featured on the American Chemical Society's website today, but they were not part of the original list which was much more descriptive (Table 9).

---

1 Albert Matlack, *Introduction to Green Chemistry* (New York: Dekker, New York, 2001).

Table 9: 12 principles of green chemistry (1998)

1. It is better to prevent waste than to treat or clean up waste after it is formed.
2. Synthetic methods should be designed to maximize the incorporation of all materials used in the process into the final product.
3. Wherever practicable, synthetic methodologies should be designed to use and generate substances that possess little or no toxicity to human health and the environment.
4. Chemical products should be designed to preserve efficacy of function while reducing toxicity.
5. The use of auxiliary substances (e.g. solvents, separation agents, etc.) should be made unnecessary wherever possible and, innocuous when used.
6. Energy requirements should be recognized for their environmental and economic impacts and should be minimized. Synthetic methods should be conducted at ambient temperature and pressure.
7. A raw material feedstock should be renewable rather than depleting wherever technically and economically practicable.
8. Unnecessary derivatization (blocking group, protection/deprotection, temporary modification of physical/chemical processes) should be avoided whenever possible.
9. Catalytic reagents (as selective as possible) are superior to stoichiometric reagents.
10. Chemical products should be designed so that at the end of their function they do not persist in the environment and break down into innocuous degradation products.
11. Analytical methodologies need to be further developed to allow for real-time, in process monitoring and control prior to the formation of hazardous substances.
12. Substances and the form of a substance used in a chemical process should be chosen so as to minimize the potential for chemical accidents, including releases, explosions, and fires.

It is important to explain that the 1998 book does not revolve around the 12 principles, but they are the subject of but one chapter. They are presented more as a mnemonic device helping to organize some ideas on what green chemistry should look like and nothing suggests that they were meant as any sort of code of conduct for chemists, and even less that Anastas and Warner saw them as an exhaustive definition of green chemistry. As already explained, some principles, for example, 6, 7, 11, and 12 had little antecedents in the works published by the EPA under the green chemistry label in the 1990s. This led to a paradoxical situation in which green chemistry was often defined as a form of environmentally friendly chemical synthesis (alongside the more traditional EPA lines), but green chemistry principles went beyond this initial ambition covering a wider range of research topics.

But what actually are green chemistry principles from an epistemological standpoint? They are neither simple laboratory guidelines nor big ethical commandments. Their status is somewhat intermediary. In the social science literature, the 12 principles are occasionally criticized for being either too narrow or, on the contrary, too broad. Both accusations have their merits. As for the first one, Roberts comments on the 12 principles in the following way:

> The Twelve Principles of Green Chemistry act to construct a boundary. The principles offer markers for those working in the field and outside of it—and help to inform others on which side of the divide they fall. The principles denote standard practices, methods, and goals. But they also clearly separate green chemistry from other types of pollution prevention. The principles work by creating a space for the practice of green chemistry and also defining and setting those practices.[2]

Elsewhere he notes that

> One can think of the writing of the twelve principles, then, in the same way one thinks of the creation of the Periodic Table. The table displays two important functions, which have led to its continued utility: it both 'accommodates' previous information while 'predicting' what future work will uncover.[3]

Roberts explains that the 12 principles set the boundaries to the field, but he warns that this is precisely their main problem, as once they become codified and incorporated into curricula in chemistry and professional trainings, they may actually stifle creativity when it comes to solving environmental problems. He notes, for example, that green chemistry practitioners insist on the broader societal challenges implied by green chemistry, and yet the 12 principles offer almost exclusively technical solutions. Roberts gives as an example the case of Edward Woodhouse. Woodhouse was a political scientist, involved in the green chemistry community already in the late 1990s, who testified before the House Science Committee at US Congress in order to safeguard additional funding for green research. He also, as explained in chapter 1, formulated the brown/green dichotomy. For Woodhouse, green chemistry had to be placed in the broader social context to fulfil its mission, as he warned against leaving the field in the hands of scientific and industrial technocrats. He envisaged green chemistry to become a tool for democratizing research policy. And yet, Roberts remarks that

> by focusing on the twelve principles alone as a way of describing/defining the practices of green chemistry, the efforts of someone like Woodhouse to address and overcome the

---

2 Jody Roberts, "Creating Green Chemistry: Discursive Strategies of a Scientific Movement" (PhD diss., Faculty of Virginia Polytechnic Institute and State University, 2006), 21.
3 Roberts, "Creating Green Chemistry," 89.

social and political obstacles facing green chemistry get excluded because they emphasize something other than the technical.[4]

Interestingly, Woodhouse himself in a socio-historical paper "Green Chemistry as Social Movement?" published in 2005 does not mention the 12 principles at all![5] It is hard not to read this omission as an explicit choice indicating that he did not embrace this way of drawing green chemistry's perimeter and that he would have wanted to make green chemistry something larger and more socially-relevant.

However, if the 12 principles are overly restrictive in their focus on technical aspects of green chemistry, they are also very permissive when it comes to what actually counts as green. The problem lies in the fact that Anastas's own position appears to be slightly inconsistent. On the one hand, as related by Roberts, Anastas said in 2004 that "the principles are not meant to be taken as "commandments," or "rules" or anything else of the sort; instead, the principles ought to be treated as "guidelines" or a 'framework for thinking.'"[6] On the other hand, Anastas wrote elsewhere that chemical research "that strives to incorporate one or more of the 12 principles" can be labelled as "excellent."[7] Historian and chemistry educator J. A. Linthorst remarked with a note of irony that if striving to incorporate only one principle of green chemistry merits the excellence label "then of course it is relatively easy to deserve the term green chemistry."[8] Following this logic, any innovation using catalysis to reduce waste, toxicity, or energy use could be considered green. For Linthorst, this user-friendliness of the term green chemistry is responsible for much of its growth as quite a few chemists may discover that their work can be actually presented as green. French economist Martino Nieddu and his colleagues went further and explained that "Anyone can then 'enter green chemistry' basing on this good conduct guide by using the few principles that they find achievable in their work."[9]

---

4   Roberts, "Creating Green Chemistry," 95.
5   Edward J. Woodhouse and Steve Breyman, "Green Chemistry as Social Movement?," *Science, Technology, & Human Values* 30 (2005): 199–222.
6   Roberts, "Creating Green Chemistry," 91.
7   Paul Anastas and Mary M. Kirchhoff, "Origins, current status, and future challenges of green chemistry," *Accounts of Chemical Research* 35 (2002): 686–694.
8   J. A. Linthorst, "An overview: origins and development of green chemistry," *Foundations of Chemistry* 12 (2010): 55–68, 67.
9   Martino Nieddu, Franck-Dominique Vivien, Estelle Garnier and Christophe Bliard, "Existe-t-il réellement un nouveau paradigme de la chimie verte?," *Natures Sciences Sociétés* 22 (2014): 103–113, 109.

Another important point to make is that in practice the principles overlap: catalysis (9) makes it possible to reduce waste (1) and energy consumption (6). The use of safer solvents (5) guarantees less hazardous chemical syntheses (3). In fact, the principles can be divided into two broad categories. Some of them offer practical solutions and others set broad goals (Table 10).

Table 10: Division of 12 principles of green chemistry according to their function

Means to an end	Goals	Somewhere in between
2, 5, 8, 9, 11	1, 3, 4, 6, 10, 12	7

Atom economy (2) is not interesting on its own, it is merely a means to reduce waste (1). At the same time, reducing derivatives (8), using catalysis (9) and real-time analysis (11) are not inherently pro-environmental. Even if a given technology satisfied all three of them, it would be absurd to claim the process is green if none of the big goals was satisfied. Principle 7 is somewhat of an outsider as it can be seen both as a guiding principle and as a practical constraint in the process design. The diverging character of these principles begs the question whether they are really the most intuitive way of expressing the green chemistry ambition. Perhaps they should be divided into two internally coherent sets of rules: large strategic principles and small practical laboratory guidelines? These inherent flaws of the codification became rapidly apparent and in the early 2000s the attempts to address them multiplied.

2. Completing and reformulating the 12 principles of green chemistry

One of the first scholars who remarked that the 12 principles of green chemistry were slightly too abstract for chemists in their everyday laboratory work was Neil Winterton from the University of Liverpool. He developed and published in 2001 the so-called "Twelve more principles of green chemistry" to better connect the practice of the laboratory and of the plant (Table 11).[10] In a brief article that appeared in *Green Chemistry*, he comments on the already existing 12 principles in the following way:

---

[10] Neil Winterton, "Twelve more green chemistry principles," *Green Chemistry* 3 (2001): G73–G75.

> Most chemists, including those working in industry, recognise these as worthwhile ideals and acknowledge that most of them already guide their work. The failure to adopt them perfectly in every case does not arise from innate perversity or from indifference to environmental considerations. It arises from technological, economic and other factors that chemists do not always address.

He then quotes William H. Glaze, the editor of the *Environmental Science & Technology* journal who said that the greenness "of a chemical transformation can only be assessed in the context of its scale-up, its application and its practice." In the following part of the article, Winterton reveal his own ambition:

> To complement and build on those formulated by Anastas and Warner and to address Glaze's concerns, twelve more green chemistry principles, objectives and requirements are suggested ... They are proposed to aid laboratory and research chemists, interested in applying green chemistry, to plan and carry out their work to include the collection of data that are of particular use to those, usually process chemists, chemical engineers and chemical technologists, wishing to assess the potential for waste minimisation.

Table 11: Twelve more principles of green chemistry (Winterton, 2001)

1. Identify and quantify by-products
2. Report conversions, selectivities and productivities
3. Establish full mass-balance for process
4. Measure catalyst and solvent losses in air and aqueous effluent
5. Investigate basic thermochemistry
6. Anticipate heat and mass transfer limitations
7. Consult a chemical or process engineer
8. Consider effect of overall process on choice of chemistry
9. Help develop and apply sustainability measures
10. Quantify and minimise use of utilities
11. Recognise where safety and waste minimisation are incompatible
12. Monitor, report and minimise laboratory waste emitted.

These new twelve principles concentrate on the methodology of the laboratory work and are much more specific than the original ones. They explain how chemists should act to achieve greenness instead of setting up goals or suggesting general technological solutions. Unlike the original principles that might be useful for teaching the foundations of the green chemistry philosophy, Winterton's principles are something to be posted as a helpful reminder in a laboratory space. And yet, in spite of their practical nature, they failed to garner any larger attention. While not completely forgotten, as they are still featured on the ACS's official website,[11] almost no major textbook mentions Winterton's list, contrary to the 1998 one which is relentlessly reproduced in every single publication on green chemistry.

Why is there such disparity? One reason may stem from the very nature of Winterton's rules: many of these principles could act as guidelines for any good chemistry, not only green chemistry. I also argue that the difference in terms of popularity between the two lists lies simply in the context of their presentation. Anastas and Warner introduced their principles in the book that is considered to be foundational for new a discipline and then they actively sought to promote them through conferences, seminars, and popular publications of the Green Chemistry Institute. Winterton's principles were expounded in a 2-page long opinion piece published before the *Green Chemistry* journal rose to prominence and there is no trace of any activism from Winterton's side to advertise his ideas.

One of Winterton's principles deserves special attention though. Principle 7: "Consult a chemical and process engineer." This is a problem often raised in the publications in these early years. Chemists, no matter how creative in their laboratories, need to be aware of the challenges of scaling-up and of the plant constraints. Another article in *Green Chemistry*, also from 2001, put it simply:

> Chemists and engineers should go for a lunch together ... [M]ost chemists tend to focus on reactions rather than the technology around the reaction and virtually all reactions are undertaken in batch reactors. Thus, in general, if a reaction does not 'work,' chemists are more inclined to change the reaction rather than investigate different equipment in which to perform the reaction. Issues of mass and energy (heat/cool) transfer, mixing, phase transfer, and general reactor design, etc., are generally not as rigorously pursued by the synthetic organic chemist as by the engineer.[12]

The idea that engineers should be more involved in making processes green was recognized by Anastas as well. Two years after Winterton's principles, in March 2003, Paul Anastas and his future wife Julie Zimmerman (working at the EPA) formulated the 12 principles of green engineering (Table 12).[13]

---

11 Booklet on various types of green chemistry principles, accessed 19/02/2022, https://www.acs.org/content/acs/en/greenchemistry/principles/design-principles-booklet.html.
12 Alan D. Curzons, David J. C. Constable, David N. Mortimera and Virginia L. Cunningham, "So you think your process is green, how do you know?—Using principles of sustainability to determine what is green–a corporate perspective," *Green Chemistry* 3 (2001): 1–6.
13 Paul Anastas and Julie Zimmerman, "Design through the Twelve Principles of Green Engineering," *Environmental Science and Technology* 37 (2003): 94A–101A.

Table 12: 12 principles of green engineering

> 1. Inherent Rather Than Circumstantial (Designers need to strive to ensure that all materials and energy inputs and outputs are as inherently nonhazardous as possible.)
> 2. Prevention Instead of Treatment (It is better to prevent waste than to treat or clean up waste after it is formed.)
> 3. Design for Separation (Separation and purification operations should be designed to minimize energy consumption and materials use.)
> 4. Maximize Efficiency (Products, processes, and systems should be designed to maximize mass, energy, space, and time efficiency.)
> 5. Output-Pulled Versus Input-Pushed (Products, processes, and systems should be "output pulled" rather than "input pushed" through the use of energy and materials.)
> 6. Conserve Complexity (Embedded entropy and complexity must be viewed as an investment when making design choices on recycle, reuse, or beneficial disposition.)
> 7. Durability Rather Than Immortality (Targeted durability, not immortality, should be a design goal.)
> 8. Meet Need, Minimize Excess (Design for unnecessary capacity or capability (e.g., "one size fits all") solutions should be considered a design flaw.)
> 9. Minimize Material Diversity (Material diversity in multicomponent products should be minimized to promote disassembly and value retention.)
> 10. Integrate Material and Energy Flows (Design of products, processes, and systems must include integration and interconnectivity with available energy and materials flows.)
> 11. Design for Commercial "Afterlife" (Products, processes, and systems should be designed for performance in a commercial "afterlife.")
> 12. Renewable Rather Than Depleting (Material and energy inputs should be renewable rather than depleting.)

It is important to note that these principles do not concern only chemical engineering but any type of creative technological work. Anastas and Zimmerman clarify that

> When dealing with design architecture—whether it is the molecular architecture required to construct chemical compounds, product architecture to create an automobile, or urban architecture to build a city—the same green engineering principles must be applicable, effective, and appropriate. Otherwise, these would not be principles but simply a list of useful techniques that have been successfully demonstrated under specific conditions.

It is clear that the 12 principles of green engineering were not just a novel set of rules for chemical engineers, in the same way the 1998 rules were directed to research chemists. They were a much more general conceptualisation with universalist ambitions. In other words, green chemistry principles could be seen as a lower-level adaptation of green engineering principles in the context of chemistry.

This direction is opposite to the one taken by Winterton who felt green chemistry principles were too vague and chemists needed more focused guidelines. Anastas and Zimmerman believed the contrary: that an all-embracing universal code for greenness was the road to take. This epistemological difference between the three sets of principles is often lost in the literature.

Now that the ambition behind the green engineering principles is visible, it is easier to understand why some of them are not complementary but simply identical to green chemistry principles. For example, principle 2 of GE [Green Engineering], concerning waste prevention, is almost identical to principle 1 of GC [Green Chemistry], and so are principle 12 of GE and principle 7 of GC (renewability). In a way, many GE principles could act as guidelines for chemists as well, but they are more far-reaching and programmatic than the GC ones.

Interestingly, one year after Anastas and Zimmerman's article (from March 2003), two scholars, Martin A. Abraham and Nhan Nguyen, published a post-conference article summarizing the outcomes of a seminar "Green Engineering: Defining the Principles" that had taken place in May 2003 in Sandestin, Florida.[14] Both Anastas and Zimmerman participated in the conference and their green engineering principles were discussed alongside the 12 principles of green chemistry and other sustainability-oriented lists such as the Hannover Principles, CERES principles, Ahwahnee Principles, and Earth Charter Principles, all developed in the 1990s. The time was ripe for similar conceptualisations in favour of sustainability. What is more interesting is that the conference did not simply 'adopt' Anastas and Zimmerman's list, but developed its own, shorter and even more general (Table 13).

---

14 Martin A. Abraham and Nhan Nguyen, "Green engineering: Defining the principles," *Environmental Progress* 22 (2004): 233–236.

Table 13: Principles of Green Engineering proposed during the Sandestin Conference in 2003

1. Engineer processes and products holistically, use systems analysis and integrate environmental impact assessment tools.
2. Conserve and improve natural ecosystems while protecting human health and well-being.
3. Use life-cycle thinking in all engineering activities.
4. Ensure that all material and energy inputs and outputs are as inherently safe and benign as possible.
5. Minimize the depletion of natural resources.
6. Strive to prevent waste.
7. Develop and apply engineering solutions while being cognizant of local geography, aspirations, and cultures.
8. Create engineering solutions beyond current or dominant technologies; improve, innovate, and invent (technologies) to achieve sustainability.
9. Actively engage communities and stakeholders in development of engineering solutions

The Sandestin principles are more society-oriented, with the most obvious example being principle 9. They also explicitly include keywords such as sustainability and life-cycle. Interestingly, since 2014, and at least as of June 2022, the English Wikipedia page on "green engineering" confuses the order in which the principles emerged.[15] It lists the 9 Sandestin principles first and then states that the American Chemical Society introduced the 12 new ones later in 2003. This is not true. Both lists were developed in 2003 and the 12 principles actually came first. Why didn't the Sandestin conference simply adopt the 12 principles of green engineering as the new code for green engineering? There is no clear answer, but perhaps lack of social dimension, as well as their rather abstract character, discouraged the wider community of engineers from simply embracing them.

It is noteworthy that both the Sandestin principles and the 12 principles of green engineering are featured on-line, as of 2022, on the official website of the American Chemical Society, alongside the 12 principles of green chemistry and the additional 12 principles of Winterton. This is interesting because, as explained before, the Sandestin principles and 12 principles of green engineering go far beyond chemistry, whereas the ACS presents them as if they were above all pertinent to chemical engineering. The problem is that these different

---

15 Wikipedia article on green engineering, accessed 01/06/2022, https://en.wikipedia.org/wiki/Green_engineering.

principles have a different epistemological status and regularly overlap, making some of them redundant. While Winterton's principles were effectively trying to complete the 12 principles of green chemistry, the 12 principles of green engineering were their generalization, and the 9 Sandestin principles were an independent codification altogether. Therefore, presenting them together makes it hard to make sense of them in a coherent manner. That is why the ACS not only offers an extensive commentary on these principles but also provides a booklet explaining how these four sets of principles relate to each other.[16]

The booklet, published in 2015, divided and reorganised them into three big categories: 1. Maximise Resource Efficiency, 2. Eliminate & Minimise Hazards and Pollution, 3. Design Systems Holistically & Using Life Cycle Thinking. This division required, however, a far-reaching interpretation and modification of the original principles to fit them into the new framework. These codifications are then 'living' documents, constantly evolving and re-explained by the institutions that deem themselves competent to do so. For example, principle 10 of green chemistry, usually described as "Design for degradation," became in the booklet "Avoid persistence," and "Safer Solvents and Auxiliaries" turned into "Find Alternatives." This is of course a nuance given that both the booklet and the ACS website provide us with more explanation. However, these small shifts are not to be neglected, especially in other reinterpretations and simplifications, since many authors tried to reshape and modify the meaning of these principles to make them more approachable and easier to disseminate.

For example, in 2005, a group of scholars, including a renowned British chemist Martyn Poliakoff, rewrote the 12 principles of green chemistry in order to fit them into the acronym "PRODUCTIVELY."[17] Three years later, the same authors added another acronym to capture the green engineering principles: "IMPROVEMENTS."[18] Thus "IMPROVEMENTS PRODUCTIVELY" were supposed to become a simple elegant way to transmit the core message of green chemistry and green chemical engineering to students and newcomers to the field (Table 14).

---

16 Booklet on various types of green chemistry principles, accessed 19/02/2022, https://www.acs.org/content/acs/en/greenchemistry/principles/design-principles-booklet.html.
17 Samantha L. Y. Tang, Richard L. Smith, and Martyn Poliakoff, "Principles of green chemistry: PRODUCTIVELY," *Green Chemistry* 7 (2005): 761–762.
18 Samantha L. Y. Tang, Richard A. Bourne, Richard L. Smith, and Martyn Poliakoff, "The 24 Principles of Green Engineering and Green Chemistry: "IMPROVEMENTS PRODUCTIVELY,"" *Green Chemistry* 10 (2008): 268–269.

Table 14: IMPROVEMENTS PRODUCTIVELY

I - Inherently non-hazardous and safe M - Minimize material diversity P - Prevention instead of treatment R - Renewable material and energy inputs O - Output-led design V - Very simple E - Efficient use of mass, energy, space & time M - Meet the need E - Easy to separate by designations N - Networks for exchange of local mass & energy T - Test the life cycle of the design S - Sustainability throughout product life cycle	P - Prevent Waste R - Renewable Materials O - Omit derivatization steps D - Degradable chemical products U - Use safe synthetic methods C - Catalytic reagents T - Temperature, Pressure ambient I - In-Process Monitoring V - Very few auxiliary substances E - E-factor, maximise feed in product L - Low toxicity of chemical products Y - Yes it's safe

These new acronyms illustrate even better the problems encountered already with the ACS booklet. Firstly, as previously explained, this type of re-writing always involves a reinterpretation of the original texts. One of the obvious examples is the letter E in PRODUCTIVELY, which refers to the E-factor. The original principles did not refer to Sheldon's E-factor but to Trost's Atom Economy (principle 2). Sheldon is much less cited in the original EPA works than Trost. His E-factor came into prominence among green chemists only in the early 2000s, and due to its popularity, it made way into these new and revised principles of green chemistry. Needless to say, Sheldon's E-factor is fully in line with green chemistry's ambitions. But it is also much larger than principle 2 that originally focused specifically on improving stoichiometric efficiency, and not on the overall process efficiency.

The same type of analysis can be applied to many other principles that were deeply transformed through these acronyms (for example V- very simple, in IMPROVEMENTS). In a certain way, one could ask whether this new rendition is not a completely new list reflecting the preoccupations of green chemists from the second half of the 2000s. How far are we allowed to modify the 12 principles of green chemistry before having to call them something else?

The second problem with IMPROVEMENTS PRODUCTIVELY is that the principles of green chemistry and green engineering were not supposed to be read together. As explained before, the 12 principles of GE are more abstract and aim to reach a wider public (all sorts of engineers). This led to a situation in which the letters R and P refer basically to the same concepts in both lists. If the authors went so far in modifying the actual content of the principles to fit them into a nice acronym, why not go a step further and remove these redundancies in an even more catchy acronym combining the two lists? It appears that the authors, in spite of their detailed exegesis of all the principles, failed to account for the fundamental epistemological difference between the two lists, and did not realize that the 12 principles of green engineering were not designed to complement the green chemistry ones, but to offer more detached and interdisciplinary criteria for greenness of all creative design activities.

The two preoccupations expressed above did not discourage a group of African scholars who, inspired by IMPROVEMENTS PRODUCTIVELY, developed in 2011 their own 13 principles of green chemistry and engineering for GREENER AFRICA, also published in the *Green Chemistry* journal.[19] This new list, as valuable as it is on its own merits, deviated even further from the original 12 principles (Table 15).

Table 15: Thirteen principles of green chemistry and engineering for GREENER AFRICA

G – Generate Wealth not Waste	A – Appropriate Materials for Functional
R – Regard for All Life & Human Health	F – Fewer Auxiliary Substances & Solvents
E – Energy from the Sun	
E – Ensure Degradability & No Hazards	R – Reactions using Catalysts
N – New Ideas & Different Thinking	I – Indigenous Renewable Feedstocks
E – Engineer for Simplicity & Practicality	C – Cleaner Air & Water
R – Recycle Whenever Possible	A – Avoid the Mistakes of Others

The GREENER AFRICA principles are broad environmental guidelines only loosely inspired by the original 1998 principles but still under their influence. Their strong point is that they avoided redundancies of IMPROVEMENTS PRODUCTIVELY and that their authors were eager to go beyond the initial framework and address the issues they found pressing. At the same time, there is here the same problem as with the original 1998 principles, namely the superposition of issues that are of a different order, for example, "Regard for All Life & Human Health" and "Reactions using Catalysts." Should these two types of preoccupations be really included in the same ruleset? Also, the letter C, "Cleaner Air & Water," expresses the old-fashioned aspiration of environmental chemistry, meaning that inventors of GREENER AFRICA were not fully satisfied with green chemistry's narrow focus on synthesis.

---

19 Nigist Asfaw, Yonas Chebude, Andinet Ejigu, Bitu B. Hurisso, Peter Licence, Richard L. Smith, Samantha L. Y. Tangb and Martyn Poliakoff, "The 13 Principles of Green Chemistry and Engineering for a Greener Africa," *Green Chemistry* 13 (2011): 1059–1060.

Overall, these lists and acronyms have all their own histories and circulate independently in different communities, but they all reinforce the central role of the original 12 principles of green chemistry from which they stem. Not only that, the idea of the 12 principles is being rehashed and reappropriated by subsequent generations of scholars. Just as much as the 12 principles of green engineering were inspired by the 12 principles of green chemistry, many more sets of green chemistry principles have been constructed in the following years. For example, in 2012 a group of French scholars formulated the 6 principles of green extraction of natural products (Table 16),[20] in 2015, Martyn Poliakoff and his collaborators wrote a programmatic article on the 12 principles of CO_2 Chemistry (Table 17),[21] in 2019, Brazilian chemists developed the 8 principles of green and sustainable education (Table 18),[22] and finally, the same year, a group of Dutch chemists laid out the 12 principles of circular chemistry to which I return in the final chapters of this book.[23]

Table 16: 6 principles of green extractions

1: Innovation by selection of varieties and use of renewable plant resources.
2: Use of alternative solvents and principally water or agro-solvents.
3: Reduce energy consumption by energy recovery and using innovative technologies.
4: Production of co-products instead of waste to include the bio- and agro-refining industry.
5: Reduce unit operations and favour safe, robust and controlled processes.
6: Aim for a non-denatured and biodegradable extract without contaminants.

---

20 Farid Chemat, Maryline Abert Vian, and Giancarlo Cravotto, "Green extraction of natural products: concept and principles," International Journal of Molecular Sciences 13 (2012): 8615–8627.
21 Martyn Poliakoff, Walter Leitner and Emilia S. Strengab, "The Twelve Principles of CO2 CHEMISTRY," Faraday Discussions 183 (2015): 9–17.
22 Vânia G. Zuin, Mateus L. Segatto, Dorai P. Zandonai, Guilherme M. Grosseli, Aylon Stahl, Karine Zanotti, Rosivania S. Andrade, "Integrating Green and Sustainable Chemistry into Undergraduate Teaching Laboratories: Closing and Assessing the Loop on the Basis of a Citrus Biorefinery Approach for the Biocircular Economy in Brazil," Journal of Chemical Education 96 (2019): 2975–2983.
23 Tom Keijer, Vincent Bakker, J. Chris Slootweg, "Circular chemistry to enable a circular economy," Nature Chemistry 11 (2019): 190–195.

Table 17: 12 Principles of CO2 CHEMISTRY

C: Catalysis is crucial
O: Origin of the CO2?
2: Tomorrow's world may be different
C: Cleaner than existing process?
H: High volume of high-value products?
E: E-factor must be low
M: Maximize integration
I: Innovative process technology
S: Sustainability is essentially
T: Thermodynamics cannot be beaten
R: Renewable (& reasonable) energy input
Y: Your enthusiasm is not enough

Table 18: 8 principles of green and sustainable education

1. Glocal problems and context
2. Investigative or research approach (openness)
3. Learning sequence and time
4. Use of surrounding resources
5. Dialogical context
6. Socio-historical context
7. Nature of content (epistemological and methodological approaches)
8. Conscientious and responsible use of water

Naturally, we can imagine a different wording. One scholar urged in his 2006 green chemistry manual for respecting the ten commandments of sustainability, but the metaphor was already out of fashion by then.[24] The choice of the 'principles of some chemistry' became more of a mnemonic device that surfs on the popularity of the original 1998 green chemistry principles, with scholars with varying backgrounds trying to promote their own codes of conduct.

---

24 Stanley E. Manahan, *Green Chemistry and ten commandments of sustainability* (Columbia: ChemChar Research, 2006).

3. Enthronement of green chemistry principles

3.1. Creating the legend

Even though the twelve principles were presented for the first time in 1998, they did not become immediately widespread. It appears that the construction of the legend surrounding the principles, and of the mythology concerning the origins of green chemistry in general, was particularly intense in the second half of the 2000s, which translated into their rapid rise in popularity in the following years (Figure 2).

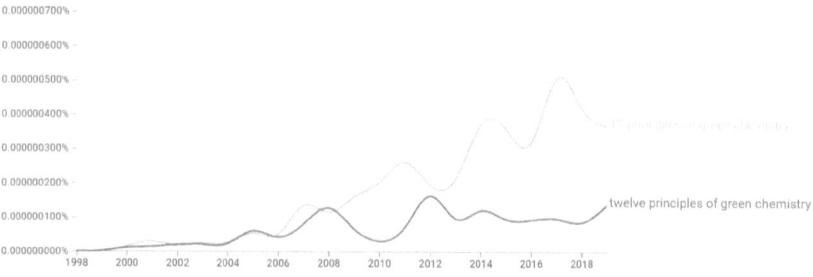

Figure 2: 12 principles of green chemistry. Popularity of the term according to Google Ngram (English-speaking books, smoothing: 0, years: 1996–2019)

In 2007, Anastas and Beach wrote:

> Since the Twelve Principles of Green Chemistry were formulated in the 1990s, there have been tremendous successes in developing new products and processes to be more compatible with human health, the environment, and sustainability goals … With the introduction of the Twelve Principles of Green Chemistry, guidelines were provided for chemists to develop clean, environmentally benign methodologies that are sustainable for the long term.[25]

The authors seem to suggest here that the 12 principles were the reason for all this favourable development over previous years. And yet, it is important to reiterate that the principles were not regularly mentioned in the *Green Chemistry* journal until around 2003 and 2004. A quick search in *Green Chemistry*'s database reveals that the term appears only once in 1999 and once in 2000, and later

---

25 Paul Anastas and Evan S. Beach, "Green chemistry: the emergence of a transformative framework," *Green Chemistry Letters and Reviews* 1 (2007): 9–24.

a few times in 2001. Furthermore, in the first half of the 2000s there were still scientific books with green chemistry in their titles that did not mention them at all.[26] Without denying the role that the principles may have played in the work of some scientists, this was far from a massive phenomenon and a tremendous amount green chemistry research was still conducted without reference to this framework. This perception was, however, changing extremely dynamically.

Already in 2006, a Swedish chemist and educator, working in the field of social studies of science, Jesper Sjöström, wrote in the pages of the *Green Chemistry* journal the following paragraph:

> In this journal, "green chemistry" is normally associated with the twelve principles of green chemistry and connected research about benign chemical reactions and products. In this article, however, I am using a broader view of green chemistry, where I include all activities and policy/knowledge areas with the aim of greening of chemistry and chemical practice. Such a broader view is useful to get an understanding of societal processes in the direction of greening of chemistry.[27]

By 2006, green chemistry became more and more defined through the 12 principles. They became a reference for understanding greenness in chemistry, and those who wanted to go further, such as Sjöström, had to actually explain that they did not mean green chemistry in the traditional sense (traditional = 12 principles). Only a few years earlier, this link had been far from established. For young scholars, such as Sjöström who got his Ph.D. in 2002, green chemistry was however inevitably seen through the lenses of Anastas and Warner's book, showing the power of their conceptualisation.

Interestingly, this was exactly the development Roberts warned against in his 2006 Ph.D. dissertation. He wrote that the solidity of the 12 principles and its success in aggregating different lines of investigation came at a price:

> First, as a didactic tool, the principles constrain and discipline their users to think about green chemistry within a certain, preordained framework. This is especially troubling in the context of a Summer School [in which Roberts participated in 2004]. Rather than playing off of the innovation and creativity of the 'next generation' of green chemists, the participants are indoctrinated into the system that the principles lay out almost immediately after arriving. The principles are also exclusionary, eliminating the participation of many through the strict technical focus of the principles. And finally, adherence

---

26 For example: Matlack, *Introduction to Green Chemistry*; Manahan, *Green Chemistry and ten commandments of sustainability*.

27 Jesper Sjöström, "Green chemistry in perspective—models for GC activities and GC policy and knowledge areas," *Green Chemistry* 8 (2006): 130–137.

and exaltation of the principles leads to unnecessary—and potentially dangerous—solidification of green chemistry as something other than chemistry in its broader sense.[28]

A few years later, even Paul Anastas realized that there were some problems with the ways that green chemistry principles had been appropriated by some scholars, who began to equate with them the entire field and perceived their formulation as the founding moment of green thinking in chemistry in general. In his later publications, Anastas often underlined that individual principles are much older than the list itself. For example, in 2016, he and his colleagues noted:

> If we look at the Twelve Principles of Green Chemistry as a structure at the time of their definition, one can see that atom economy and atom utilization had already been proposed in 1990, the efforts in biodegradability were well established, waste minimization or 'clean' reactions were discussed, utilization of renewable feedstocks had been in place for decades, catalysis for lower energy and selectivity was a perennial focus in industry and academia, and accident prevention through inherent safety had been well-recognized.[29]

The same article states that

> After 25 years, the 12 Principles are no longer perceived as a collection of elements, they have been transformed into a cohesive system. This shift in thinking has been accompanied by a transformation of Green Chemistry from a way of getting rid of bad things (inefficient syntheses, toxic reagents, needless waste, etc.) into an approach that is used as a way to generate better things (in essence a tool for innovation to generate new performance, function, and efficiencies).

And then a remarkable phrase follows: "Obviously, they do not provide a simple "check list," where a chemical reaction or process will automatically get "greener" with the number of principles addressed in its development."

And yet, in spite of this reassurance, the principles, perhaps victims of their own success, became a checklist. In fact, in many cases, the authors recognize their limits, insist on their holistic character, and then immediately fall into a familiar ground of enumerating the principles and associated discoveries. This is particularly visible in the discussions on green chemistry education, as well as in more conceptual papers discussing the achievements of green chemistry as a field (or the discipline's 'self-presentation').

---

28 Roberts, "Creating Green Chemistry," 96.
29 Paul Anastas, Buxing Han, Walter Leitner, Walter Leitner and Martyn Poliakoff, "'Happy silver anniversary:' Green Chemistry at 25," *Green Chemistry* 18 (2016): 12–13.

3.2. 12 principles in green chemistry education

The case of education is perhaps the most sensitive since there is an entire generation of chemists that are being educated into the green chemistry principles as if it was the sole way of conceiving greenness in chemistry. Without specifically targeting any particular approach, I want to present a few recent examples of green chemistry education papers that fall victim to this approach.

In one article from 2019, the authors describe their cross-disciplinary course curriculum (art and chemistry) and its evaluation methods. There are, for example, the following activities:

> Discuss how some of the 12 principles of green chemistry are applied in this moss graffiti activity. Address a minimum of three green chemistry principles ... List two principles of green chemistry and explain how these two principles were applied in the unit on environmentally friendly paints. ... For all paint mixtures you synthesized and used, you should address the following: Identification of which of the 12 principles of green chemistry are applied in this experiment and justification.[30]

In a different paper from 2015, its authors describe their own class in which they asked students to

> Compare and contrast the two C-N bond-forming methods with respect to **3 of the 12 principles** [bold in original] of green chemistry discussed in class (choose the 3 that you feel are most relevant to this step).[31]

In both cases, the green chemistry principles are treated as if they were a checklist to tick if a given reaction/process satisfies them. Such a checklist approach was probably most famously formalized by a Portuguese team that proposed in 2010 a novel type of green chemistry metric: Green Star, which was meant to become an important tool for teaching green chemistry at the university level.[32]

---

30 Anne Marteel-Parrish and Heather Harvey, "Applying the principles of green chemistry in art: design of a cross-disciplinary course about 'art in the Anthropocene: greener art through greener chemistry,'" *Green Chemistry Letters and Reviews* 12 (2019): 147–160, 154.

31 Alexander E. Waked, Karl Z. Demmans, Rachel F. Hems, Laura M. Reyes, Ian Mallov, Erika Daley, Laura B. Hoch, Melanie L. Mastronardi, Brian J. De La Franier, Nadine Borduas, and Andrew P. Dicks, "The Green Chemistry Initiative's Contributions to Education at the University of Toronto and Beyond," *Green Chemistry Letters and Reviews* 12 (2019): 187–195, see supplemental information page 17.

32 M. Gabriela T.C. Ribeiro, Dominique A. Costa & Adélio A.S.C. Machado, "'Green Star:' a holistic Green Chemistry metric for evaluation of teaching laboratory experiments," *Green Chemistry Letters and Reviews* 3 (2010): 149–159.

> The basic idea of GS [Green Star] is to construct a star with a number of corners equal to the number of principles used for the evaluation of the synthesis reaction, all the 12 or only some if the remaining are not applicable, each corner with length proportional to the degree of accomplishment of the corresponding principle – a semi-quantitative view of the global greenness of the reaction can then be obtained by looking at the star and appreciating its area: the larger the area, the greener is the reaction. … The construction of the metric consists in evaluating the greenness of the reaction for each principle (in a scale from 1 to 3, maximum value of greenness).

This is perhaps the most direct translation of the 12 principles into a practical tool for greenness evaluation. The authors offer precise criteria for attributing a note, ranging from 1 to 3, to every single principle, conducting at the same time their rather detailed exegesis. Of course, one can immediately see many problems with the Green Star approach. Does evaluating the toxicity or biodegradability of a given process on a scale from 1 to 3 really convey the complexity of the issues at stake? For example, the three stages attributed to biodegradability (principle 10) by the authors are 1) biodegradable, 2) may be rendered biodegradable, and 3) non-biodegradable. While some simplification is unavoidable, one has to ask at what point the simplification renders the evaluation meaningless.

The Green Star and sister concepts (Green Circle and slightly different Green Matrix) are rarely cited in the literature on green chemistry metrics, despite the fact they were initially conceived to address the mass metrics deficiencies (Atom Economy and E-Factor). They appear, however, in the literature on green chemistry education and are sometimes recommended as a didactic tool.[33] This is not surprising since this type of evaluation is easy to convey and easy to apply. As a consequence, these concepts influence the way young chemists perceive green chemistry. The students are being educated into the 12 principles as both guiding principles and genuine criteria for judging the greenness of a product or a reaction.

---

33 Laura B. Armstrong, Mariana C. Rivas, Michelle C. Douskey, Anne M. Baranger, "Teaching Students the Complexity of Green Chemistry and Assessing Growth in Attitudes and Understanding," *Current Opinion in Green and Sustainable Chemistry* 13 (2018): 61–67; M. Gabriela T. C. Ribeiro, Santiago F. Yunes, and Adélio A. S. C. Machado, "Assessing the Greenness of Chemical Reactions in the Laboratory Using Updated Holistic Graphic Metrics Based on the Globally Harmonized System of Classification and Labeling of Chemicals," *Journal of Chemical Education* 91 (2014): 1901–1908; Zuin, Segatto, Zandonai, Grosseli, Stahl, Zanotti, and Andrade, "Integrating Green and Sustainable Chemistry into Undergraduate Teaching Laboratories."

The point is not to criticize the papers mentioned above, as they all present a range of fascinating ideas worthy of learning from, but they are emblematic of an entire category of educational approaches in which students are expected to use the 12 green chemistry principles as a check-list; a philosophy against which Anastas explicitly warned. In the first two cases mentioned above, a minimum of three principles is mentioned. Is three enough for a process to be green? But what about the fact that some principles are goals and other means to achieve them? Since it does not make sense to call a catalysis green if it does not aim to reduce waste or energy consumption, it will always accompany other principles. But then, if it does, are two principles (e.g. catalysis and waste reduction) not sufficient on their own to label a process as green? And what if some principles are satisfied but others grossly violated? And what if a reaction or a process has an overall positive environmental impact but does not fall into any of the principles? These problems are of course well understood by the practitioners of green chemistry, but such a check-list approach is not conducive for transmitting the complexity of sustainability problems to students who discover the environmental challenges in chemistry for the first time. This approach is reductive and not holistic.

The underlying question concerning all these approaches is why to even base any evaluation on the green chemistry principles and not on, for example, the green engineering ones. What gives the 12 principles of green chemistry such a privileged status in contrast to other lists also recommended by the ACS? They are certainly very popular because their history is intertwined with the history of the concept of green chemistry itself, but nothing suggests in the 1998 book that they should be seen as anything more than helpful guidelines, one of many others. The explanation lies in the fact that they became seen by the community of green chemists as the ultimate summary of what green chemistry is supposed to be.

3.3. Green chemistry's self-representation

In chapter 1, I mentioned ChemisTree, a tree-like visualization of progress and of contributions to green chemistry presented through the 12 principles. The image was originally published in 2018 in the *Green Chemistry* journal to celebrate the 20th anniversary of the codification. The authors, sixteen of them in total including Anastas, lay down in the article, principle by principle, various technologies, studies, and solutions associated with every single of the 12 principles. The aspiration was to create a comprehensive overview of the discipline. Why the tree metaphor?

> The maturity of the field also inspired us to introduce the "Green ChemisTREE" metaphor ... Tree diagrams have been used in chemistry to celebrate the diversity of applications that can be supported by a particular raw material, ... for example tracing a product back through its polymer, monomer, or other intermediate components and ultimately a resource such as crude oil, coal, natural gas, or minerals. Indeed, many variations on the theme have appeared in illustrations since the beginnings of modern chemistry ... Our goal is similar: to be concise, informative, visual, and encourage the viewer to reflect on what lies at the roots of progress in the chemical enterprise.[34]

From the historical point of view, however, the use of a tree metaphor is misleading. Of course, the authors are right to say that this metaphor has been long used to show the diversity of products that can be extracted from oil, and it certainly can be used as a metaphor to show how an idea, a theory, or a method can bring about novel methods, theories, and so on. Nonetheless, this reading of the recent history of green chemistry is completely incorrect. The 12 principles are more like a bundle of concepts tied together on purpose by people theorizing about what green chemistry should look like, but they have all different genealogies and often developed independently. Creating a sophisticated metaphor in which green chemistry is a trunk, principles are branches, and leaves are specific methods and technologies creates a false sense of unity. It is not a good history of science and, arguably, it is not a good representation of green chemistry either.

The authors visibly struggle with the metaphor themselves in the body of the text because some principles overlap and different chapters have to constantly tie one to another. That is why the section on principle 1 (waste prevention) has to anticipate the content of the section on principle 2 (atom economy) and of many other sections. Moreover, some sections are extremely broad, whereas others focus on very narrow technological improvements. At one point the authors write:

> Principle 3 calls upon chemists to consider factors other than efficiency when evaluating environmental and social impacts. It is closely connected to Principle 4, which focuses on design of less hazardous products, and also takes into account the properties of precursors, side products, and waste.

This is of course correct, but one could ask why these two principles could not have simply been merged together. Is it because the authors remain prisoners of the 1998 classification that had been hastily constructed as a mnemonic tool, and not as an ultimate guide to chemical greenness?

---

34 Hanno C. Erythropel et al., "The Green ChemisTREE: 20 years after taking root with 1 the 12 Principles," *Green Chemistry* 20 (2018): 1929–1961.

Another problem with illustrating the history of green chemistry through the 12 principles is that many technologies and competencies mentioned there have little relation with the research actually conducted by people identifying with green chemistry. This is visible many times in the text, especially in the description of more autonomous principles such as 7 (renewability) and 11 (monitoring of chemical reaction). The latter is particularly striking since the article mentions developments in sensors, chromatography, various types of spectroscopy, and computing. None of these belongs, on its own, to green chemistry in any meaningful way (unlike developments in pollution prevention strictly speaking) and the authors have to constantly tie principle 11 to other principles and justify the inclusion of a given technology in their list. The tree metaphor may suggest that the invention and the development of these various techniques was related in some way to the history of green chemistry, and this is the message that young chemists discovering the field may take from the article, even though in many cases this is evidently not true.

The purpose of this criticism is not to undermine the utility of such an overview in general, and even less the quality of the research conducted by numerous authors of the ChemisTree paper. On the contrary, the usefulness of such a comprehensive overview of topics relevant for green chemistry leaves no doubt, but the theoretical framework has to be clearly delineated. Should such a paper present topics studied by the green chemistry community and developed with its theoretical tools, or all topics that are considered relevant to the green chemistry practice? These two approaches would refer to different bodies of literature. The ChemisTree paper does not address this question at all. Another pressing question is, of course, whether the division of all green chemistry-related innovations into 12 separate principles that have different scope, character, and different genealogies, is the most efficient way to tell the history of the field? Finally, if the 12 principles are not a check-list and should be evaluated holistically, isn't this kind of analysis of principles, one by one in different chapters, actually against the spirit of the codification itself?

Conclusions

In 2021, the editor-in-chief of an important review *Green Chemistry Letters and Reviews*, prof. Anja-Verena Mudring, wrote

> There can be no doubt that our civilization needs chemistry for a prosperous and healthy life. In this, the '12 Principles of Green Chemistry' were truly a game-changer. We learned as professionals that with a thoughtful upfront design it is possible to provide

new chemicals, products and technologies that are safe for people and environment, and that we are able to take good stewardship of our planet.

She assured immediately after that

> The concept of Green Chemistry does not stand against economy. Green Chemistry has inspired many innovative R&D activities and chemical industry around the World has adopted its principles. Green Chemistry has sparked innovation, new ideas and progress. Green Chemistry is not a static field. It grows with the challenges we are meeting – climate change and associated catastrophic phenomena; persistent environmental pollution from plastic and microplastic, materials that seem to be essential for packaging, automobiles, consumer electronics, building materials, health care products and more; and an ongoing pandemic.[35]

An overly simplistic and slightly caricatural reading of this editorial may suggest that thanks to the game-changing 12 principles (that have been, I remind, barely known until the mid-2000s in the wider community of chemists), humanity is now equipped to respond to the climate change and the coronavirus pandemic, and all of this, the shareholders rest reassured, without challenging our economic model. This reading is obviously unfair and I do not imply that this is exactly what prof. Mudring had in mind. And yet, there is a thin line between pitching one's own discipline and educating generations of young chemists into a certain paradigmatic narrative they might not be able to critically assess. The 12 principles have become an unquestionable reference, impossible to avoid in any discussion on what green chemistry is and should be. Do they deserve this place? Were the principles of green engineering not more comprehensive and more programmatic? Were Winterton's principles not more practical? Isn't there a risk that a strict adherence to the green chemistry principles may discourage some researchers from following less-trodden paths?

For the historian, what matters is that after having been intensively promoted by Anastas and his close circles for years through programmatic papers and conferences, the principles gained sufficient inertia to snowball on their own and dominate the language of people practising green chemistry. And yet, their popularization did not cut debates on another fundamental question: what exactly counts as green?

---

35 Anja-Verena Mudring, "Editorial," *Green Chemistry Letters and Reviews* 14 (2021): 1.

Chapter 4: What is green chemistry? (normative approach)

A quick glance at the growing number of publications in journals devoted to green chemistry may give the impression that a tremendous amount of research is being currently conducted on environmentally friendly processes and products. However, how do we know whether they are genuinely green? It seems obvious that if someone wants to publish in *Green Chemistry* and other similar venues, the editors and reviewers should not exclusively rely on the authors' self-assessment, but should evaluate the greenness claims themselves. After all, declaring something 'green' is a powerful political statement. International chemical companies claim to care about the environment because 'being green' in the eyes of the consumers translates into companies' profits. Green claims are also an element of national and international politics as governments boast that they fund environmentally friendly research projects and advocate sustainable policies to secure re-election.

The problem is that since the outset of the discipline, the criteria for evaluating whether given products and processes were green or not were fuzzy at best. Of course, the 12 principles became a benchmark of sorts with time, but the vast majority of scholars have been aware of their limitations and knew that things may occasionally get difficult to interpret. In this chapter, I show that establishing the criteria of greenness is very far from straightforward.

1. Major controversies: chlorine sunset and fracking

A 2002 editorial in *Green Chemistry* written by the biochemist James K. Bashkin offers some interesting insights on the boundaries of the discipline:

> Green chemistry is so diverse that we often have trouble communicating amongst ourselves, and often spend too much time arguing about what is really Green. When it comes to definitions of the field, there are purists and there are revisionists; there are the dogmatic and the lackadaisical.[1]

He then makes an important observation:

---

1 James K. Bashkin, "Green Chemistry: Can we rally together, or will we fragment into pieces?," *Green Chemistry* 4 (2002): G14.

> When it comes to chemistry, there are those who do it and those who talk about it, those who set policy and fund grants, and those who spend money generously provided by government and private funding agencies.

It is crucial to remember that definitions vary between these various stakeholders and the expectations about greenness can largely differ depending on who they come from. Therefore, the role of argumentative strategies and theoretical battles over greenness can deeply influence the direction taken in practice by researchers.

Bashkin offers then his own understanding of greenness:

> any chemical advance or engineering advance that significantly reduces the burden placed on the environment by current industrial processes is helping to clean up the environment, by definition. Many of these advances may not be fully Green: there may still be some waste generated, there may still be lifecycle issues that have not fully been considered, and there may still be unforeseen consequences that this new chemistry presents to the environment. We must be as vigilant as possible to avoid doing any greater harm while we attack the obvious problem processes with new vision. ... I would like to see similar principles applied in the evaluation of academic work on the subject of Green chemistry. In many cases, we will have to take small steps towards the ultimate greening of a particular chemical reaction. As long as these steps move the chemistry in the right direction, I believe they should be heard by the community and collected in this journal.

This is a classic 'gut-feeling' green chemistry. If it feels that a process or a technology is good for the environment and that the advancement counterbalances possible side effects (e.g. considerably less waste, even if slightly more toxic), it merits being considered green. Bashkin subscribed to the idea that small steps in the right direction were the best way to proceed. Green chemistry is in his view above all 'the chemistry of greening.' This is a perfectly justifiable stance, much in line with what Anastas himself said in the 1990s, and a tremendous amount of valuable environmentally friendly research has been conducted by scientists with this attitude. This is arguably the dominant way of thinking in the field.

What are the consequences of this philosophy for the publication practice? Following Bashkin's logic, it may be argued that even if the reviewer of a given paper does not entirely agree with the greenness claim concerning a given process or reaction, or if he or she finds the improvement minimal, the article should be published anyway as it may potentially, in the grand scheme of things, contribute to the debate. This philosophy admits that further studies may disprove the greenness claims by, for example, a life-cycle analysis. This sounds right at first. But then, are such analyses conducted on a systematic basis on the articles published in *Green Chemistry* in the past? Can we be sure that the products, reactions, and processes described in its pages ten years ago remain genuinely green today? Of course, it is perfectly normal that some older results are disproven by the progress of science, but this is done through clear-cut empirical and laboratory research. However, the problem of greenness criteria is much more multi-layered and value-laden. What if an article published in Green Chemistry loses its greenness status due to the change in the definition of what counts as green or not? Should it be retracted? Adjusted?

As explained above, the greenness claims may have political consequences and the presence of an article in a green chemistry journal is a powerful statement. This is even more true when it comes to major environmental controversies. The public may expect green chemistry to be particularly useful in meaningfully engaging with these topics and contributing to the debate. And yet, this is where the shortcomings of intuitive greenness criteria are the most visible.

One of the earliest examples of such a failure concerns the chlorine controversy. The 1980s and 1990s was a period of fierce debates over the use of chlorine and chlorine-based chemicals. Some advocated the so-called 'chlorine sunset,' a process of phasing out chlorine derivatives from every-day and industrial uses. The conflict opposed environmental activists, among others from Greenpeace, and a number of scientists and engineers often close to the industry. The topic was discussed in the pages of scientific journals such as *Nature* (which rather openly sided with environmentalists)[2] and galvanized first blogs and social networks in the early day of the internet. In 2000, the famous book *Pandora's Poison: Chlorine, Health, and a New Environmental Strategy* became a scientific bestseller.[3] The traces of this debate are still to be found on-line with pro-chlorine websites violently attacking environmentalists and Rachel Carson's *Silent Spring*.[4]

In this tense period, exactly in 2000, the *Green Chemistry* journal published Neil Winterton's article "Chlorine: the only green element – towards a wider acceptance of its role in natural cycles."[5] Winterton was, as already explained, the author of the 12 additional principles of green chemistry. I have already

---

2 Terry Collins, "A call for a chlorine sunset," *Nature* 406 (2000): 17–18.
3 Joe Thornton, *Pandora's Poison, Chlorine, Health, and a New Environmental Strategy* (Cambridge, MA: MIT Press, 2000).
4 See for example Michael Fumento's "Rachel's Folly" from 1996, accessed 20/02/2022, http://lobby.la.psu.edu/015_Disinfectant_Byproducts/Organizational_Statements/C3/C3_The_End_of_Chlorine.htm.
   Fumento, a conservative pundit, made himself recently known for denying the existence of the Covid-19 epidemics on his personal website, https://www.fumento.com//.
5 Neil Winterton, "Chlorine: the only green element – towards a wider acceptance of its role in natural cycles," *Green Chemistry* 2 (2000): 173–225.

mentioned his article in the context of its ambivalent attitude towards Rachel Carson's heritage, but the article itself constitutes above all a thorough and comprehensive overview of chlorine compounds and their circulation in the natural environment. Winterton admits that to scientists,

> particularly those who have been concerned with the environmental impact of some persistent, bioaccumulative and toxic compounds containing chlorine, the title [of the article] may appear flippant, provocative or even heretical. Yet others, aware of my former industrial affiliation, might search for some ulterior motive in the sub text of my survey.

While the paper is not explicitly political, Winterton takes nevertheless sides in the debate. He says, for example, that

> while water purification and disinfection has been ranked 46th by Life Magazine in its top one hundred advances of the millennium, concerns arising from epidemiological evidence, suggesting a statistically weak association between the consumption of chlorinated drinking water and liver cancer, led the authorities in Peru, presuming a causal link, to stop chlorinating drinking water. The cholera epidemic that ensued killed more than 10 000 people. Fortunately, more rational judgments about the balance of benefits arising from the use of chlorine are beginning to appear.

Whether chlorine is a "green element" or the "Pandora's poison" can be a matter of discussion, but what counts is that early green chemists parted ways with environmentalist movements over the chlorine sunset or, the bottom line, they failed to take a definitive stand on the issue as a community, as no comment or rebuttal of Winterton's article was ever published in the journal's pages. Green chemistry was not the chemistry of environmentalist movements, to be sure.

But perhaps this attitude can be blamed on the lack of tools in these early days? After all, the 12 principles were not yet widespread at that time. Perhaps they could help to solve the riddle of chlorine and allow green chemists to formulate some common ground (even if Winterton himself knew the 12 principles, but did not consider them in his chlorine article). Unfortunately, Jody Roberts notes that during the Green Chemistry Summer School he attended in 2004, the issue of chlorine sunset was raised and no agreement was reached either. On the contrary, when one of the participants

> strongly criticized the lack of support given by green chemists for the enactment of [chlorine sunset] policy ... Paul Anastas asked [him] whether he had considered the fact that those advancing green chemistry had purposefully decided not to engage in

the debate because it didn't mesh with the strategy being employed by the leadership of green chemistry. ... Satisfied with the lack of an answer, Anastas left the room.[6]

For Roberts, the green chemists failed in this case to prove themselves to be a genuine pro-environmental movement. However, I argue that this can be also read as an epistemological failure of the green chemistry concept itself. For the framework that, on the one hand, roots its mythological story in the activism of Rachel Carson and the early environmentalism and, on the other hand, incorporates ideas such as life-cycle assessment and sustainability, its incapacity to at least offer some tools to refine the debate on key environmental controversies, such as the chlorine sunset, is disquieting.

The chlorine sunset is only the tip of the iceberg of similar controversies on which green chemistry, as a framework with a distinct methodology, failed to bring any meaningful answers. For example, in 2002, Clark struggled with the problem whether we can "consider chemical transformations and materials processing that involve radiation as 'green?'" (he was inclined to do so).[7] A much more egregious example comes however from 2015, when an introductory editorial commented on an article submitted to *Green Chemistry*, which had aroused many questions in the editorial team. The topic of the controversial article was the improvement of fracking technology.[8]

> Fracking is a very controversial technique for which a number of potential hazards to the environment are discussed. Most significantly, one may raise the fundamental question of whether an increased exploitation of fossil resources is inherently incompatible with the Principles of Green Chemistry. To be honest, we were unable to find a consensus and a final answer to this question ourselves. Thus, we turned to the classical maxim 'in dubio pro reo.'

As much as the green chemistry community was unable to build a consensus over chlorine, it appears that there was no consensus on fracking either. And from this lack of consensus, the editors inferred that the article could get published in the leading green chemistry journal. Things get even more interesting when confronted with the justification that followed:

---

6 Jody Roberts, "Creating Green Chemistry: Discursive Strategies of a Scientific Movement" (PhD diss., Faculty of Virginia Polytechnic Institute and State University, 2006), 140.
7 James Clark, "Green chemistry initiatives in Japan," *Green chemistry* 4 (2002): G54.
8 Walter Leitner, "The subject of 'fracking' in Green Chemistry," *Green Chemistry*, 17 (2015): 2609.

> I have often used the phrase "If Sustainability is your goal, Green Chemistry is the way." Is unconventional oil and gas recovery a sustainable technology? Certainly not in the long term – but it is and will be for quite some time influencing the basis of our raw material and energetic value chain, having an enormous impact on many aspects of our environment today.

To put it differently, the author claims that sustainability is the criterion for defining what green chemistry should strive for, then admits that fracking is not sustainable, but then decides that the article satisfies the criteria of green chemistry anyway. Clearly not at ease with the chosen solution, he continues to try to justify the decision:

> Does this mean that we will now all at a sudden encourage the green chemistry community to focus on improvements of existing technologies, even if they are only incremental steps aside into the right direction on otherwise clearly unsustainable paths? Certainly not: we need to continue our efforts to contribute with fundamentally new approaches to a sustainable chemical industry and lay the basis for disruptive green technologies. However, we also recognize that things are not always black or white and there are more than 50 shades of green. In this particular case, we have decided to bring the topic on the table and to shed some light on the chemistry that is involved in fracking from the green chemistry perspective.

One could scoff that some of these 50 shades of green are surprisingly close to brown. The issue is not whether fracking can be made more environmentally friendly, perhaps it can, but this does not mean the topic has to be presented in the leading green chemistry journal. It may have been a perfect fit for a journal on mining-related questions. A publication in a green chemistry journal is a value-laden statement, 'look, our technology is green,' a strategy easily exploitable by companies trying to green-wash their activities. However, in the paragraph above there is one more curious statement. The first phrase states that the green chemistry community should not merely focus on an incremental approach to greening already existing technologies. However, this was exactly what Bashkin argued for in the *Green Chemistry* journal 13 years prior! It was all about small steps, not groundbreaking innovations. This divergence of opinion may be read as the evolution of the journal's editorial line, but it can also indicate an uncomfortable truth: throughout its entire existence, green chemistry struggled to define its own identity and the editors of the *Green Chemistry* journal long appeared to accept that greenness lies in the eye of the beholder.

2. Core problem: ionic liquids

Some could retort that chlorine or fracking are heavily politicized issues and one cannot judge an entire discipline (community? paradigm?) through their lenses. However, even some of the core topics forming part of green chemistry's identity were put into question over the years. This is the case of ionic liquids.

Principle 5 states that "The use of auxiliary substances (e.g. solvents, separation agents, etc.) should be made unnecessary wherever possible and innocuous when used." This is a pretty straightforward rule encouraging the use of solventless methods or innocuous solvents. What are innocuous solvents? The literature on green chemistry usually cites water, supercritical CO_2, and, for example, the so-called ionic liquids. The ionic liquids are mentioned, although only once, in Anastas and Warner's 1998 book as well as in Clark's introductory article editorial to the first issue of the *Green Chemistry* journal from 1999. Shortly afterwards, however, they became one of the leading topics in the green chemistry community. If we look at the most popular keywords from green chemistry journals, ionic liquids always reach the top ten.[9] In 2002 was published a voluminous work *Ionic Liquids, Industrial Applications for Green Chemistry* which is the second most cited book with green chemistry in its title (after Anastas and Warner's book) according to Scopus.[10] Three years later, in 2005, prof. Robin Rogers got the Presidential Green Chemistry Award for his work on "A Platform Strategy Using Ionic Liquids to Dissolve and Process Cellulose for Advanced New Materials."[11] What are ionic liquids?

> A working definition suitable for a historical view of ionic liquids, and perhaps a practical view, is that an ionic liquid is a salt with a melting temperature below the boiling point of water. Most salts identified in the literature as ionic liquids are liquid at room temperature, and often to substantially lower temperatures

clarifies an enthusiastic article from *Green Chemistry* published in 2002.[12] The article introduces an entire issue devoted to these curious substances. Why are ionic liquids green? The same article explains:

---

9   The challenges of the bibliometric analysis are presented in Chapter 5 .
10   Robin D. Rogers and Kenneth R. Seddon, *Ionic Liquids, Industrial Applications for Green Chemistry*. (Oxford: Oxford University Press, 2002).
11   Presidential Green Chemistry Award 2005, accessed 20/02/2022, https://www.epa.gov/greenchemistry/presidential-green-chemistry-challenge-2005-academic-award.
12   John S. Wilkes, "A short history of ionic liquids—from molten salts to neoteric solvents," *Green Chemistry* 4 (2002): 73–80.

> The link between ionic liquids and green chemistry is clearly related to the solvent properties of ionic liquids. Most of the papers in this volume will be somehow related to ionic liquids as solvents, and the one property that stands out above all others is the huge liquidus range. Liquidus range is the span of temperatures between the freezing point and boiling point of a liquid. The consequence for green chemistry is that ionic liquids are the ultimate non-volatile organic solvent, with emphasis on the 'non-'. No molecular solvent (other than molten polymers) comes even close to the low volatility of ionic liquids.

Are then ionic liquids the paramount of green chemistry in the early 2000s? Things are a little bit more complicated. Over the period, more and more chemists signalled problems with this category of solvents. For example, a 2003 article in *Green Chemistry* warned in its title: "Ionic liquids are not always green" and then pointed out their potential toxicity.[13] In 2005, a news article entitled: "Warning Shot for Green Chemists: Some Solvents with an Environmentally Friendly Reputation May Kill Fish" appeared in *Nature*.[14] The article refers to ionic liquids specifically in the context of green chemistry; they are the solvents that green chemists are known to endorse. Roberts commented on the affair in 2006:

> Not only are many of [these] solvents toxic, they are potentially toxic at far smaller quantities than traditional solvents. Thus the resistance faced by ionic liquids and attempts to include them in the field continues to shape the space and practice of green chemistry.

He also noted that back in 2004, even though ionic liquids had been one of the core topics of the green chemistry summer school, there had been also many voices of concern about whether they were green at all.

It is, as usual, important to add nuance to the message. What I am saying here is that some uses of ionic liquids can, certainly, improve the greenness of a given reaction or process. This is a perfectly valid hypothesis that can be proven through meticulous life-cycle analyses alongside the life-cycle analyses of the solvents they aim to replace. The problem is that ionic liquids were viewed as green by default. They were a tool for achieving the greenness label.

This shortsightedness concerning ionic liquids has been noted by many chemists in the following years. In 2009, Maggel Deetlefs and Kenneth R. Seddon, some of the most recognized specialists in the field of ionic liquids, wrote:

---

13 Richard P. Swatloski, John D. Holbrey, and Robin D. Rogers, "Ionic liquids are not always green: hydrolysis of 1-butyl-3-methylimidazolium hexafluorophosphate," *Green Chemistry* 5 (2003): 361–363.

14 Mark Peplow, "Warning Shot for Green Chemists: Some Solvents with an Environmentally Friendly Reputation May Kill Fish," *Nature* (3 November 2005), https://doi.org/10.1038/news051031-8.

Although ionic liquids still uphold their baptism as 'green', the use of the term, and indeed what constitutes a green ionic liquid synthesis, continues to be misinterpreted. At the heart of the misinterpretation, and many spurious claims, lies the assumption that ionic liquids are green because of their negligible vapour pressure. However, there are many other factors that determine whether an ionic liquid is, or is not, green, particularly the ionic liquid preparation itself... In other words, the term 'green' should only be applied to an ionic liquid if both the ionic liquid, and the process used to produce it are green, i.e. if all twelve principles of green chemistry apply.[15]

All twelve principles! This is an extremely high bar for any process to satisfy, but the authors later give considerably more margin by stating that the publications on ionic liquids in *Green Chemistry* should at least indicate which principles are satisfied. Their entire article is a thorough analysis of various ways to prepare ionic liquids, but it concludes with a somewhat disgruntling statement: "we believe this is the first critical assessment of the greenness of the synthetic procedures and purification methodologies commonly used for the synthesis of ionic liquids." If the first such assessment was published in late 2009, does it mean that the greenness of the preparation of ionic liquids had never been systematically evaluated before? The problem is not that new discoveries weakened the greenness claims of a given substance. That would be the normal way the science proceeds. The problem is the superficiality of the overall greenness framework that allowed for this claim to be unsubstantiated for years.

In 2011, another editorial recognized that in the early 2000s, the topic of ionic liquids had exploded in the pages of *Green Chemistry*.

In those earlier days the claim that ionic liquids were green solvents largely rested on the (then believed) non-volatility of ionic liquids and its associated properties (low flammability, ease of containment etc.). This claim has since been challenged on several occasions, particularly with the toxicity and environmental persistence of the most widely used ionic liquids being noted as important negative green factors.[16]

More importantly, however, the article explained the evolution of the journal's editorial line, i.e. what would count as green.

Today we would consider the notion of any solvent being intrinsically green to be based upon too narrow a view of what it means to be green. As a discipline, we would now judge a process solvent in terms of the overall environmental impact of the process that it is a part of.

---

15 Maggel Deetlefs and Kenneth R. Seddon, "Assessing the greenness of some typical laboratory ionic liquid preparations," *Green Chemistry* 12 (2010): 17–30.
16 Tom Welton, "Ionic liquids in Green Chemistry," *Green Chemistry* 13 (2011): 225.

At the same time, the article admits that nothing is intrinsically green. "If the use of a more hazardous solvent gave significant improvements to the total environmental impact of the overall process than a less hazardous alternative, then the greener choice is the former." The editorial underlined that ionic liquids were still welcome in the pages of the journal, but not in any context, in the sense that an article on the polarity of an ionic liquid, even if potentially interesting to some green chemistry practitioners, would not have a place in the journal any more. On the other hand, only articles that "lead to environmental benefits" were to be accepted. And yet, the editorial specified, "That is not to say that these articles must always be at the near application stage of the research, with full life-cycle analysis completed, but an environmental benefit must be the purpose of the work." The purpose is, however, a vague criterion based on self-assessment.

In a comprehensive overview "Searching for green solvents" published by the future Editorial Board Chair of the *Green Chemistry* journal, Philip Jessop, in 2011, the author insisted that there was indeed a problem in the green chemistry community:

> Too many "green" solvents have been hyped as being green without the claim being verified in any way. In particular, single issue arguments have been proposed for justifying a claim of greenness, such as "it is green because it is biomass-derived." Even when solvents have been assessed on more than one parameter, it is most customary to assess the environmental risk posed only by the solvent itself and not the chemistry that was required to synthesize the solvent.[17]

Jessop's contribution provides guidelines for chemists interested in establishing and evaluating the greenness of their solvents, but it is also interesting because of a small sociological and scientometric analysis he conducted. In fact, he presents in the article two diagrams. The first is the result of a survey among chemists asking the question "what class of solvents will be responsible for the greatest reduction in environmental damage?" The results place 1) CO_2 as the most popular answer, followed by, 2) water, 3) selected organic solvents, 4) ionic liquids, and others in subsequent positions. The second pie chart shows the proportion of papers describing each class of solvent published in *Green Chemistry* in 2010. The results are the following: 1) ionic liquids (almost half of all of the articles), 2) water, 3) solventless, 4) CO_2, and others. Surprised, the author asks: "Are we studying the right solvents?" And continues elsewhere:

---

17 Philip G. Jessop, "Searching for green solvents," *Green Chemistry* 13 (2011): 1391.

the research currently being performed in the academic green solvents community is not aimed at the applications that are most likely to lead to a reduction in environmental damage. Because we are focusing on minor applications of solvents, the environmental benefits from our research are likely to be minor.

Not only can ionic liquids be potentially environmentally troublesome, not only did the assessment of their greenness remain very subjective and intuitive, but most notably, they were not even seen as particularly promising according to the green chemists themselves. It looks like there was a persistent mismatch between what green chemists considered to be green, and what they actually practised. Of course, the field of ionic liquids should not be dismissed entirely and many introductory textbooks fully recognize their limitations and advantages.[18] But the problem concerning the evaluation of their greenness remains.

At this point, it should become more than clear that the unfortunate story of ionic liquids is the direct result of the framework for green chemistry that has been chosen in the 1990s. Ionic liquids appear to satisfy principle 4 in chemical synthesis. If, hypothetically, green chemistry had been from the outset focused on toxicological consequences, as advocated by Garrett, or if the pollution prevention had remained closely tied to pollution remediation, so in line with Hancock's environmental chemistry, the ionic liquids might have never gained such popularity. It is not the negligence or shortsightedness of chemists, but the limitation of the framework that is to blame.

3. Green chemistry metrics

A different way to deal with the problem of evaluating greenness is to construct methods and tools for quantifying and measuring it. Quantification would allow us to rapidly compare different chemistries and judge which one is greener. I have already discussed this quantitative approach in chapter 2 giving the two examples of Barry Trost's Atom Economy and Roger Sheldon's E-Factor. Today, these methods are known collectively as green metrics. The term itself became widespread relatively late. The concept of green metrics (in chemistry) was occasionally mentioned in the early green chemistry handbooks, but it was in the second half of the 2000s when it really took off. Sheldon's 2007 paper "The E Factor: fifteen years on" and the books such as the 2008 *Green chemistry metrics. Measuring and Monitoring Sustainable Processes* sparked a lot of interest and led

---

18 See for example: Peter Wasserscheid, Annegret Stark, Paul Anastas, ed., *Green Solvents: Ionic Liquids*, (Wiley-VCH, 2013), 2–3.

to the emergence of a plethora of further publications.[19] Green chemistry metrics is, without doubt, one of the most promising and interesting fields of chemical research, operating with a distinct language that makes an autonomous field and not just a part of green chemistry in general.

What led to its development in the late 2000s? Everything indicates that it was a collective answer to the problems with defining what should be considered green and what should not. There are some revealing passages in Sheldon's 2007 article. Sheldon summarizes what he thinks about green metrics in a paragraph entitled "E Factor and atom efficiency as green metrics:"

> It is now generally accepted that two useful measures of the (potential) environmental acceptability of chemical processes are the E factor, defined as the mass ratio of waste to desired product, and the atom efficiency, calculated by dividing the molecular weight of the desired product by the sum of the molecular weights of all substances produced in the stoichiometric equation.[20]

A few paragraphs later Sheldon completes:

> All of the metrics discussed above take only the mass of waste generated into account. However, what is important is the environmental impact of this waste, not just its amount, i.e., the nature of the waste must be considered. One kg of sodium chloride is obviously not equivalent to one kg of a chromium salt. Hence, we introduced the term 'environmental quotient,' EQ, obtained by multiplying the E factor with an arbitrarily assigned unfriendliness quotient, Q. For example, one could arbitrarily assign a Q value of 1 to NaCl and, say, 100–1000 to a heavy metal salt, such as chromium, depending on its toxicity, ease of recycling, etc. The magnitude of Q is obviously debatable and difficult to quantify.

Let us recapitulate. By 2007, according to Sheldon, the two principal green chemistry metrics were Atom Economy (or efficiency as he calls it) and E-Factor. However, in practice, they measured exclusively one of the 12 principles: principle 2, concerning the mass of the waste produced in a given process. In spite of the all-embracing character of the 12 principles, the green chemistry community did not generalize by 2007 any other meaningful green metric. Sheldon's environmental quotient, as important as it is, is an extremely subjective rudimentary metric. It is almost surprising that toxicity and environmental impact of

---

19 Mike Lancaster, *Green chemistry: an introductory text* (Cambridge: Royal Society of Chemistry, 2002), 69; Roger A. Sheldon, "The E Factor: fifteen years on," *Green Chemistry* 9 (2007): 1273–1283; Alexei Lapkin and David J. C. Constable, ed., *Green chemistry metrics. Measuring and Monitoring Sustainable Processes* (London: Wiley-Blackwell, 2008).
20 Sheldon, "The E Factor: fifteen years on," 1274.

processes and products were not quantified in a more formalized manner in the field that aspired to solve the problem of toxic waste from its outset.

Equally interesting is the fact that while Sheldon affirmed in 2007 that E-Factor and Atom Economy were generally accepted green metrics, only ten years later both of them fell from grace. For example, a recently (2018) published comprehensive volume with the simple name *Green Metrics* comments on the past metrics such as Atom Economy in the following way: "Atom economy is mentioned here for the sake of its historical place, but as a metric to drive green chemistry or for greening a reaction, it is not a particularly practical, informative, or useful metric."[21] And that is not even because of the lack of ecotoxicological evaluation but because it does not "correlate well with mass, energy, or waste associated with chemical reactions." As for Sheldon's E-Factor the book points out a different problem:

> While the metric has been tremendously helpful, it may in practice be subject to a lack of clarity depending on how waste is defined by the user. … For example, is only waste from one manufacturing plant in view, and is waste from emissions treatment … included? Is waste from energy production … included? Or perhaps something like waste solvent passed onto a waste handler and burned in a cement kiln is not included.

The authors correctly remark that the notion of waste is a fickle one, and this qualification depends on various economic factors, a problem that has also been studied in environmental history.[22]

I will not enter into any more details as it would require lengthy explanations on different metrics, a topic that goes far beyond the scope of the book. The core message of this section is that the entire field of green chemistry metrics is very young and began flourishing only in the 2010s. Out of 118 articles published in *Green Chemistry* between 1999 and 2018 containing the keyword "green metrics," only 8 were published before 2010, and the vast majority after 2015. It does not mean, however, that the expansion of this field solves the problem of the definition of greenness. On the contrary, even if progress is being made, some scholars express frustration over the lack of coherence in the vocabulary concerning green chemistry metrics. In a lengthy passage, one scholar explained the problem in the following way in 2018:

---

21 David J.C. Constable and Concepción Jiménez-González, *Green Metrics* (Weinheim: Wiley-VCH, 2018), 12.
22 Marcin Krasnodębski, "The Social Construction of Pine Forest Wastes in Southwestern France During the Nineteenth and Twentieth Centuries," *Environment and History* 28 (2022): 155–183.

> Another problematic issue is the coining of apparently "new" metrics which are simple arithmetic manipulations of existing metrics. For example, Clark recently introduced waste intensity (WI) and optimum efficiency (OE) as "new" metrics when in fact they are simply the ratio of E-factor to process mass intensity (PMI) and the ratio of global reaction mass efficiency (gRME) to atom economy (AE), respectively. To some these extra terms may be interpreted as redundant and therefore do not add any real value to the discussion of parameterizing the degree of greenness for a given reaction or synthesis plan to a desired product. Both trends of re-branding existing names with new ones and manipulating established metrics into new ones are expected to continue for the foreseeable future in step with the noted trend of divergence. ...
>
> At present, even after two decades of study, there is still no universal acceptance of any single or set of metrics that best describes material efficiency for an individual reaction or synthesis plan. Academics tend to gravitate to overall yield and number of steps in a synthesis plan, while workers in the pharmaceutical industry have embraced process mass intensity as the preferred metric along with its variants.[23]

In other words, the core debates continue to focus on ways of measuring the amount of waste. Leaving ecotoxicology aside, issues such as the use of renewable materials or the process safety (principle 11) are for the most part not taken into account in these metrics, no matter how sophisticated. Some scholars, dissatisfied with the mass-centred approach, embraced a more radical green chemistry orthodoxy by strictly adhering to the original founding texts and to the 12 principles. That was the case of the Portuguese team behind the Green Star I mentioned in the previous chapter. They explicitly wrote in 2012 that the Green Star was developed due to the insufficiencies of mass metrics.[24]

Others, however, opted for a more heterodox, or perhaps even syncretic, approach, and embraced conceptual frameworks coming from sustainability studies. This is, of course, not a new idea, the sustainability and metrics were discussed already in 2001,[25] but it was in the 2010s when the tendency really

---

23 John Andraos, "Useful Tools for the Next Quarter Century of Green Chemistry Practice – A Dictionary of Terms and a Dataset of Parameters for High Value Industrial Commodity Chemicals," *ACS Sustainable Chemistry and Engineering* 6 (2018): 3206–3214.

24 Gabriela T.C. Ribeiro and Adélio A.S.C. Machado, "Greenness of chemical reactions – limitations of mass metrics," *Green Chemistry Letters and Reviews* 6 (2013): 1–18.

25 Alan D. Curzons, David J. C. Constable, David N. Mortimera and Virginia L. Cunningham, "So you think your process is green, how do you know?—Using principles of sustainability to determine what is green–a corporate perspective," *Green Chemistry* 3 (2001): 1–6.

took off with the popularization of the life-cycle assessment (LCA) and more sophisticated concepts as a part of green chemistry itself.

Conclusions: Redrawing the boundaries

In 2017, a long-time green chemistry scholar David Constable offered a small retrospective on the history of green chemistry boundaries.

> When I first became interested in green chemistry in the mid-1990's I would attend green chemistry related conferences or read the green chemistry literature. While listening or reading I was bothered by the assertions that what was being presented was green. On reflection, the only reason I could see that the research was green was because the author or presenter made the assertion that it was green. For example, a total synthesis of a new chemical entity was being presented. Step 8 of the 12 step synthesis was calculated to have a high atom economy. The remaining 11 steps were characterized by low yields, poor atom economy, large stoichiometric excesses of reagents and reactants, high mass and energy intensities (low efficiencies), solvents, reactants and reagents which had a variety of major environmental, health and safety hazards ... When questions were raised about these other ways of evaluating whether or not a synthesis was green, the questions were usually dismissed as being irrelevant and/or highly misguided in light of the amazing chemistry that was presented.[26]

Since the 1990s, the processes and evaluation methods were perfected and the questions about greenness are not scolded any more. And yet, the self-assessment of whether a given process is green is still commonplace and the editorial committees still struggle with the question of where to draw a line concerning greenness. What is certainly clear, is that this line has been redrawn many times. An article publishable in *Green Chemistry* in 2010, might not have been in 2020.

In 2020, Philip Jessop gave probably the most comprehensive overview of the editorial policy of the *Green Chemistry* journal to date:

> The vast majority of papers submitted to the journal Green Chemistry describe a new chemical, reaction, product or process. Any author who writes such a paper believes, or at least hopes, that their discovery is greener than the previous best, but simply claiming that something is green is insufficient. The readers of today are more skeptical and demand more evidence. In fact, the journal Green Chemistry has a specific policy of requiring such evidence. The journal website states: To be published, work must present excellent science and a significant advance in green chemistry. Papers must contain a comparison with existing methods and demonstrate advantages over those methods

---

26 David J.C. Constable, "The practice of chemistry still needs to change," *Current Opinion in Green and Sustainable Chemistry* 7 (2017): 60–62.

> before publication can be considered. ... How can an author of a paper compare their new chemistry to the prior art? Greenness is much more difficult to measure than the rate of a catalyzed reaction, and is a task that most chemists have never been trained to do.[27]

Jessop follows by enumerating four options, from worst to the best, to compare different chemistries. The bottom line is the comparison by a single factor (1). "Pick one parameter, such as mass of waste or energy consumption, and quantify it for the new chemistry and the previous best chemistry." This is the least satisfying approach, discouraged, albeit acceptable if alternatives are harder to achieve. The next step is a qualitative comparison by several factors (2). Jessop enumerates categories such as: mass of waste, human toxicity, carcinogenicity, smog formation, or water consumption. Yet a better option is life cycle thinking (3), so any attempt to incorporate a more global way of thinking about a given process or product. The ideal article is the one that includes a full-blown life-cycle Assessment (4). Obviously, not every LCA is equal to another and they can themselves incorporate a range of different factors, but Jessop points out that "if you or a collaborator can supply an LCA in the paper or in an accompanying paper, then you will have given more evidence for a green advance than 99% of the papers ever published in the field of green chemistry."

It is interesting to note that today, the life-cycle assessment, which was developed outside of the green chemistry movement, became the most desirable criterion for evaluating whether something is green or not. To an extent, this constitutes a failure of the green chemistry theoretical framework that did not manage to elaborate greenness criteria on its own. Should the green chemistry principles not indicate precisely those procedures, techniques, and methods that enable to reduce the need for long and expensive LCAs? This plea was explicitly formulated by Anastas in his 1996 introductory chapter on green chemistry.[28] Green chemistry was supposed to provide the way of making things environmentally friendly even without the LCA.

But of course, the LCA itself is a multi-layered activity, with many shades of green and its own methodological problems. By making the LCA the optimum criterion for evaluating greenness, all green chemists should be aware of these

---

27 Philip Jessop, "Editorial: Evidence of a significant advance in green chemistry," *Green Chemistry* 22 (2020): 13–15.
28 Paul Anastas and Tracy C. Williamson, "Green Chemistry: An Overview," in *Green Chemistry. Designing Chemistry for the Environment*, ed. Paul Anastas and Tracy C. Williamson (Washington D.C.: American Chemical Society, 1996), 1–17, 3.

Conclusions: Redrawing the boundaries

internal challenges and limitations. More importantly, if only a fraction of LCAs are comprehensive enough to include all key environment- and sustainability-related factors, and 99% of green chemistry papers do not include any LCA at all, then really well-documented green chemistries are extremely few and far between.

The main problem lies in the fact that *Green Chemistry* is the flagship journal of the field and its experts are among the most aware of the challenges of green evaluation. Its editors and reviewers know that they should require proof of green claims, so what we find in it is the best-documented green chemistry there is. However, green chemistry grew exponentially with many other journals appearing throughout the 2000s and 2010s. Of course, in the pages of some of them, there are similar questions being pondered. Klaus Kümmerer, one of the leading German chemists in the field and the creator of another important journal established in 2016, *Current Opinion in Green and Sustainable Chemistry*, wondered in 2017:

> Can you already classify a product as green if just one of the Twelve Principles ... is fulfilled, for example, by using renewable feedstocks? Or must all Twelve Principles be fulfilled? How should they be weighted against each other? And what if a chemical or material meets all the Twelve Principles but is simply not needed?[29]

These types of fundamental questions still gave chemists sleepless nights in the late 2010s, supposedly 25 years since the emergence of the field.

However, the articles using the tag or keyword "green chemistry" are published in numerous traditional chemical journals, purely based on the self-assessment of the authors. The editors and reviewers of these journals may be perfectly competent in evaluating good or bad chemistry, but are they all aware of debates on the criteria of greenness evaluation? Do these articles that 'identify' as green really belong to the green chemistry paradigm/discipline? The popularity of the term did not make its frontiers any sharper. On the contrary, green chemistry appears to be more heterogeneous than ever.

---

29 Klaus Kümmerer, "Sustainable Chemistry: A Future Guiding Principle," *Angewandte Chemie – International Edition* 56 (2017): 16420–16421.

Chapter 5: What is green chemistry? (descriptive approach)

In the previous chapter, I explored a range of normative attempts to define green chemistry. I studied the opinions of people on what green chemistry should be as well as methods supposed to evaluate greenness. However, definitions are one thing and practice is another. The alternative to a normative approach would be a descriptive approach — a study of what has been actually published and presented as green chemistry in scientific publications over the last few decades. This type of analysis can give a good understanding of the reception of the concept among practitioners of chemistry and allow us to evaluate its success. It can show us what typical scientists think that green chemistry really is. For the years 1999–2020, the Scopus database returns more than 30 000 publications with the term "green chemistry" in their titles, abstracts, keywords, or in the name of the journal. A full-text search in a much more comprehensive but less precise Google Scholar reveals more than half a million references to green chemistry. The number of new green chemistry articles published every single year has been growing in an exponential manner almost doubling every five years and the end of the trend is nowhere to be seen (Figure 3). Every single day dozens of new articles tagged that way are being indexed.

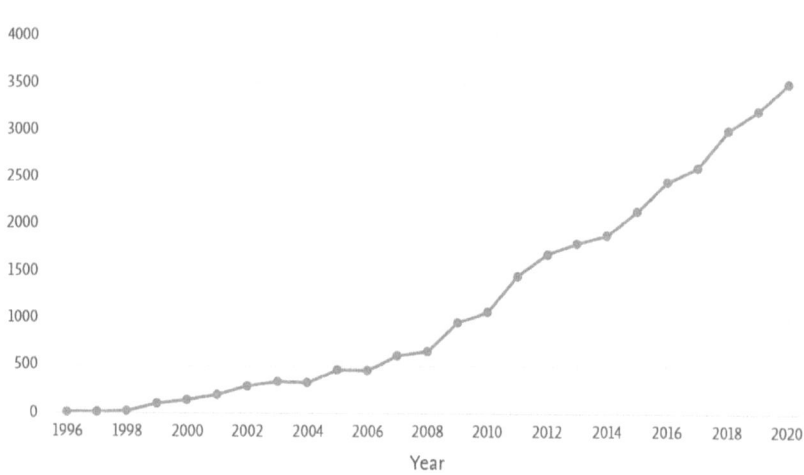

Figure 3: Number of green chemistry articles published yearly (1996–2020) according to Scopus[1]

1. Methodology

I analyse the history of green chemistry research through five 5-year periods (1996–2000, 2001–2005 2006–2010, 2011–2015, 2016–2020), characterising every single one of them with a few parameters, for example, the most prolific authors, the most cited articles, and the most popular keywords. I adapt parameters to every period in order to highlight only the most relevant developments. This is then not a comprehensive detailed bibliometric analysis, but more of an attempt to tell a story through numbers. The reason for dividing the history of green chemistry into separate time chunks results from the fact the field was not practised the same way in these different periods. If we conducted a scientometric analysis of the entire body of literature in green chemistry, the results would be skewed towards the recent fashions, simply because the number of publications is growing every year.

All the bibliometric analyses are established with the Scopus database and concern exclusively the documents that are properly indexed. In Scopus there is also the category "Others documents" in which there are mentions of books not directly indexed in Scopus, but cited in the documents in the database. For the sake of consistency, I do not include them in the analysis and mention them

---

1 See the methodology section for the details on how the number was calculated.

only if relevant. The choice of Scopus over Web of Science (and Google Scholar) results from the ease of use of its tools for scientometric studies. Both Scopus and WoS cover largely the same publications so discrepancies should be minimal. Google Scholar is considerably larger, but data are harder to extract and interpret. Of course, this means that many publications remain under the radar in the study that follows, but since the goal is not to represent exact numbers but rather to indicate broad tendencies, this approach should be perfectly sufficient.

The bigger problem concerns the fact that Scopus tags articles not only with the keywords explicitly provided by the authors but also with the so-called "indexed keywords" provided by content suppliers and then standardized.[2] In practice, it means that keywords such as "Article," "Conference Paper," "Chemical Reaction," or "Chemistry" are consistently returned with any search. These keywords constitute a useless noise and should be eliminated from the analysis. While these are easy to dismiss, others, such as "Unclassified Drug," "Environmental Impact," or "Chemical Industry" are more problematic. Their use is extremely inconsistent between similar articles and for this reason I tend not to include them in this study at all, even if they may be useful in a different search. Another problem is that some keywords are very similar, for example, "Catalysts," "Catalyst," and "Catalysis," as well as "Synthesis" and "Synthesis (Chemical)." Again, some articles are tagged with more than one of these keywords, whereas others are not. I seek to aggregate these different categories into a more readable format. Overall, readers should be warned that the keyword analysis in this chapter required a lot of interpretation to construct comparable data and is the most objectionable part of the entire book. As a consequence, while I indicate the most popular keywords in every period, the results should be approached with care. Further studies may add nuance to them. In order to avert the problem, I offer a complementary analysis of the most cited articles in every 5-year period to show the general tendencies in publication trends.

Another question that can be asked is whether the chronological, and not thematic, presentation is the best choice. While some changes would be more visible in the comparison theme by theme, I argue that the chronological approach allows us to systematically comment on the evolution taken by green chemistry and slowly uncover how the field changed.

The most challenging problem is, however, different. What kind of search terms should we input into the databases? The starting point is, of course, the

---

2 Scopus website support hub, accessed 21/02/2022, https://service.elsevier.com/app/answers/detail/a_id/21730/supporthub/scopus/.

term "green chemistry." The typical search in Scopus and Web of Science concerns titles, abstracts, and keywords. However, this choice has some pitfalls. In 2010, J. A. Linthorst published one of the few comprehensive articles on the history of green chemistry in which he presented a plot showing publication trends in green chemistry based on Web of Science. He obtained a result according to which the share of the *Green Chemistry* journal in the total number of articles featuring the term "green chemistry" in titles, abstracts, and keywords fell from 25% in 1999 to around 5% in 2006.[3] The message he tried to convey was that green chemistry rapidly spread to a wide range of different journals and the flagship *Green Chemistry* was responsible for but a fraction of publications in the field. The problem with this analysis is that authors publishing in *Green Chemistry* may not feel the need to indicate the term "green chemistry" anywhere in their articles. After all, they publish in a journal with the term in its official name, why mention it then in the title or the keywords! On the contrary, they may be discouraged from repeating it. The result of this choice is that Linthorst largely underestimated the importance of *Green Chemistry* in shaping the discourse of the discipline, even though the overall tendency was more or less correct.

To get a more accurate picture of the field, one must include all the articles published in journals with "green chemistry" in their names as well. This includes of course *Green Chemistry* (created in 1999) but also Royal Chemical Society's book series *RCS Green Chemistry* (in the editing of which both Clark and Anastas played a prominent role) as well as *Green Chemistry Letters and Reviews* (created in 2007), and some much more recent reviews. Since our objective is to follow the term "green chemistry," I decided to include one more journal with "green" in its name: *Current Opinion in Green and Sustainable Chemistry* (2016), in spite of the fact that it has a distinct publication philosophy. At the same time, another important journal, *ChemSusChem*, was not included in my analysis since its scope is, in theory, broader than green chemistry, therefore the authors publishing there may be more inclined to tag their articles as green chemistry if they feel their study makes part of it. The goal of this perhaps somewhat artificial division is to avoid as much as possible conflating sustainable and green chemistry (this is, of course, impossible with *Current Opinion in Green and Sustainable Chemistry*). Whether this ambition is justified or not is discussed in Chapter 7. However, even if *ChemSusChem* was included, as well as the term "sustainable chemistry" in our search terms, the results would be only barely affected since

---

3   J. A. Linthorst, "An overview: origins and development of green chemistry," *Foundations of Chemistry* 12 (2010): 55–68, 63.

green chemistry is a far more popular term. As a consequence, my Scopus query concerns 1) the term "green chemistry" in titles, abstracts, and keywords and 2) all the publications in the journals and book series with "green chemistry" and "green and sustainable chemistry" in their titles.

A similar search strategy was adopted, for example, in a recent comprehensive and sophisticated bibliometric analysis entitled "Intellectual authorities and hubs of Green Chemistry."[4] This admirable study relied on data extracted from Web of Science instead of Scopus and offered numerous insights into the respective 'weight' and influence of green chemistry's key protagonists. It identified six major thematic clusters in green chemistry (solvents, ionic liquids, biomass, catalysis, Green Chemistry characterization, and CO2 as substrate) and tried to capture the networks of authors working on these topics. My chapter is considerably simpler from a methodological standpoint and focuses on highlighting much more general trends in the ways the term green chemistry was used. While this may be seen as less 'scientific' and cruder, especially in the light of scientometrics as a discipline, it also allows for a more exploratory and open-minded approach that does not study green chemistry as if it was a clearly delineated object, which appears to be the case in the paper above.

To reiterate, in the sections that follow, for the sake of clarity and to avoid overcharging the text, for every 5 year-period I present only the figures relevant to the narrative to indicate shifts and major developments. It means that some sections may be better illustrated than others not because of the lack of data but because the data do not show anything particularly interesting. Overall, all graphs in this chapter, except for the keyword analysis that had to be cleaned to be presentable, are reproducible by every Scopus user. This may, however, change if Scopus decides to tag more older articles with the green chemistry keyword or, perhaps as a result of evolving fashions, it switches the tag to sustainable chemistry or something else. As such the results are valid as of June 2021. As explained in the introduction, this does not invalidate them but on the contrary, preserves the internet landscapes for future historians.

2. Birth of the discipline (1996–2000)

I have already presented in chapter 2 a more qualitative study of the formative years of green chemistry, but a scientometric analysis of this period provides

---

[4] Leonardo Victor Marcelino, Adilson Luiz Pinto, and Carlos Alberto Marques, "Intellectual authorities and hubs of Green Chemistry," *TransInformação* 32 (2020).

us with a handful of new observations against the backdrop of which it will be possible to compare subsequent developments. It should be reminded the *Green Chemistry* journal was established in 1999 and I include all the papers published in its pages in the analysis. The consequences of this choice can be observed in the list of the most productive authors from the period (Figure 4).

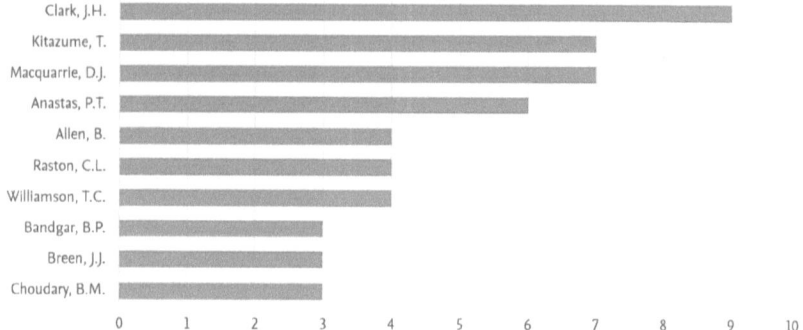

Figure 4: Most productive authors writing about green chemistry in 1996–2000 (number of articles)

Clark is the leader of the list with no less than 9 articles, but Kitazume (Tokyo), and Macquarrie (Clark's collaborator from York), who occupy the second and third place respectively, published numerous articles in the pages of *Green Chemistry* as well. Just behind the podium, there is Paul Anastas with 6 articles indexed in Scopus. Among other authors there are, for example, Tracy Williamson with whom worked Paul Anastas, and the first director of the Green Chemistry Institute, Joseph Breen. Overall, two poles, American and British, emerge through this reading.

What immediately strikes us is of course the low number of articles. This can be partly explained by the fact that not all articles from these early years are indexed in Scopus but undoubtedly the field was clearly in its infancy. Interestingly, between 1996 and 2000, out of 253 articles indexed, 138, so roughly half, came from the *Green Chemistry* journal. It means that from the very beginning there was a substantial number of publications referring to and identifying as green chemistry published elsewhere. They were, however, largely dispersed, as illustrated in Figure 5.

Birth of the discipline (1996–2000)   143

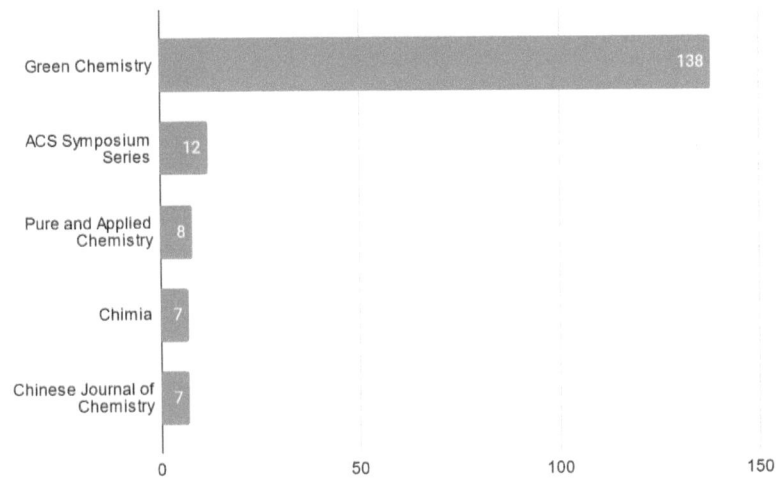

Figure 5: Five most popular venues publishing green chemistry articles (1996–2000)

Behind *Green Chemistry* with 138 articles (all articles published in the journal), there is *ACS Symposium Series* in which seminal papers from 1994 and 1996 were published as well as huge international journals such as *Pure and Applied Chemistry* and *Chimia*, but they include only a handful of papers, meaning that they were followed but dozens of journals in which the term appeared sporadically. It is important to reiterate though, that a substantial number of papers are missing in this analysis because they were not properly indexed in the database.

As for the national representation, as already established, the US and the UK were clearly dominant, but India, Japan, and China were close followers (Figure 6). A very low number of articles from Germany, as compared to the size of the German chemical academia and the importance of sustainability in German research, is noteworthy and I explain it in the chapter on sustainable chemistry.

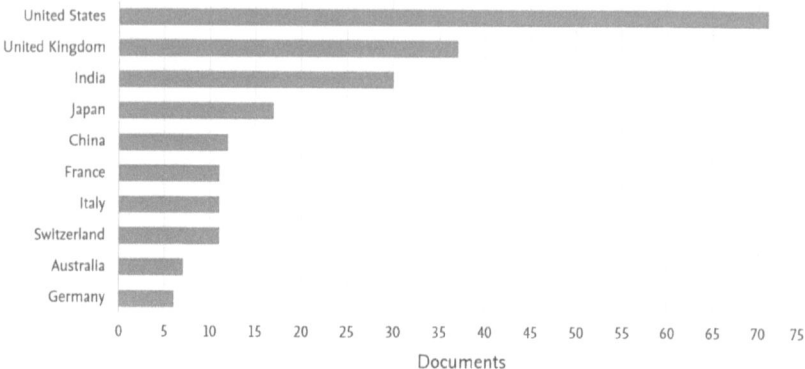

Figure 6: Green chemistry articles by national affiliation of the authors' institution (1996–2000)

What were the most popular topics studied in these first five years? This is without doubt the hardest problem to probe, as explained in the introductory part. Figure 7 attempts to offer a glimpse at some of the most frequently appearing 'meaningful' keywords used in green chemistry articles after cleaning up the data.

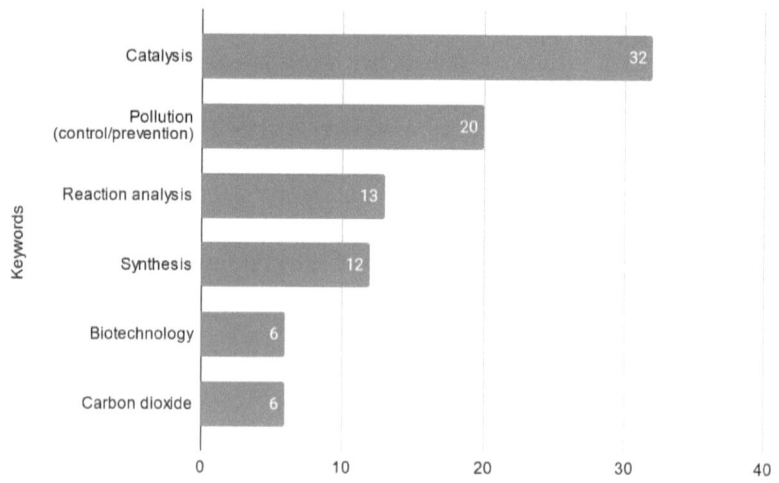

Figure 7: Most popular keywords in green chemistry papers (1996–2000)

The exact numbers should be taken with a grain of salt. I combined, for example, "catalysis" with "catalyst" and "homogenous catalysis" (some of these may be redundant). Also, it appears that many articles in *Green Chemistry* were not properly tagged at all (the numbers are largely inferior to the total number of publications) and, for example, the topic of ionic liquids that played a major role in *Green Chemistry* at the time is not adequately visible in the analysis. Still, some takeaways are possible. For example, the primacy of catalysis as a tool for achieving greenness, as well as the high count of articles tagged as pollution control/prevention. The importance of the latter could be probably increased if keywords such as "environmental monitoring" were counted as well. Due to these imperfect measures, one more analysis of the publications should be conducted to evaluate what the most important and influential topics were. Table 19 illustrates the most cited articles published between 1996 and 2000 (as of 1st of June 2021).

Table 19: Most cited green chemistry articles published between 1996–2000 (in bold papers published in *Green Chemistry*; citation count as of 1st of June 2021)

	Title	Year	Citation count
1	**Solvent-free organic syntheses. Using supported reagents and microwave irradiation**[5]	1999	1293
2	**Green Chemistry: Challenges and opportunities**[6]	1999	697
3	**Diels-Alder reactions in ionic liquids**[7]	1999	514
4	Synthetic pathways and processes in green chemistry.[8]	2000	410

---

5 Rajender S. Varma, "Solvent-free organic syntheses. Using supported reagents and microwave irradiation," *Green Chemistry* 1 (1999): 43–55.
6 James Clark, "Green Chemistry: Challenges and opportunities," *Green Chemistry* 1 (1999): 1–8.
7 Martyn J. Earle, Paul B. McCormac and Kenneth R. Seddon, "Diels–Alder reactions in ionic liquids. A safe recyclable alternative to lithium perchlorate–diethyl ether mixtures," *Green Chemistry* 1 (1999): 23–25.
8 Pietro Tundo, Paul Anastas, David StC. Black et al., "Synthetic pathways and processes in green chemistry. Introductory overview," *Pure and Applied Chemistry* 72 (2000): 1207–1228.

| 5 | Catalysis of liquid phase organic reactions[9] | 1998 | 406 |

Out of the five most cited articles, three are scientific strictly speaking (1, 3, 5) and confirm our intuitions about the most popular topics over the period. Solventless reactions, ionic liquids, and catalysis were all what at the core of green chemistry back then. The two remaining papers are more theoretical and programmatic and they explore the concept of green chemistry in general. The second article was written by Clark and the fourth article was co-authored by Anastas. Clark's short article is the first editorial to his journal published in 1999 introducing green chemistry to the wider scientific public. Article 4 is, on the other hand, a more comprehensive overview of the topic stemming from a collaborative project led by IUPAC's work group on green chemistry, which fully adopted the EPA's understanding of the field. Interestingly, Anastas's paper is the only one on the list that actually connects to the EPA's framework and to the 12 principles. None of the other papers relied on any theoretical framework of greenness, which was understood in a very intuitive way, and their authors did not seem to be aware of Anastas's conceptualisations. If one wanted to define green chemistry as the conscious result of the work within the framework provided by the 12 principles, four out of five most cited articles in green chemistry between 1996 and 2000 did not satisfy this criterion at all.

Why is Anastas and Warner's book not included in the overview above? As I explained in the methodology part, the reason is that Scopus does not index many older books. It does, however, count the citations of non-indexed documents and enables limited analyses of the so-called "other documents." I do not conduct a detailed study of these publications, but if we look at the entire period we are studying (1996–2020), we obtain the following result (Table 20).

---

9   James Clark and Duncan J. Macquarrie, "Catalysis of liquid phase organic reactions using chemically modified mesoporous inorganic solids," *Chemical Communications* 8 (1998): 853–860.

Table 20: Most cited "other documents" in green chemistry (published between 1996 and 2020; citation count as of 1st of June 2021)

Authors	Title	Year	Citation count
Anastas, P. T., Warner, J. C.	Green Chemistry: Theory and Practice	1998	7309
Rogers, R. D., Seddon, K. R.	Ionic Liquids, Industrial Applications for Green Chemistry	2002	919
Anastas, P. T., Williamson, T. C.	Green Chemistry: Frontiers in Benign Chemical Syntheses and Processes	1999	794
Clark, J., Macquarrie, D.	Handbook of Green Chemistry and Technology	2002	589
Lancaster, M. 2002	Green Chemistry: An Introductory Text	2002	477

The first five most cited documents non-indexed in Scopus (all of them are books, no articles) were published between 1998 and 2002. The bias towards the older publications is understandable considering the fact that later books were often properly included in the database. What can be immediately observed is the massive advantage of the 1998 foundational book. With more than 7000 citations it is by far the most cited green chemistry document ever. The second book is an overview on the topic of ionic liquids, a proof of this subject's importance for green chemists' identity. The third place is occupied by another book edited by Anastas, similar in scope to the 1996 ACS publication I dissected in chapter 2. The two following books come from Britain. Clark's book is a more advanced work, whereas Lancaster's publication is more student-friendly. Overall, these five books shaped the way green chemistry was understood by generations of scholars.

3. Explosion of interest (2001–2005)

The number of green chemistry articles published every year doubled between 2001 and 2005, as the field continued its staggering growth. A total of 1573 articles published between 2001–2005 are indexed in Scopus, with almost half of them (715) in *Green Chemistry*. In these years green chemistry is still a niche but becomes increasingly visible outside the initial community (Figure 8). Let us also note that in this period, there are no other journals devoted exclusively to green chemistry yet.

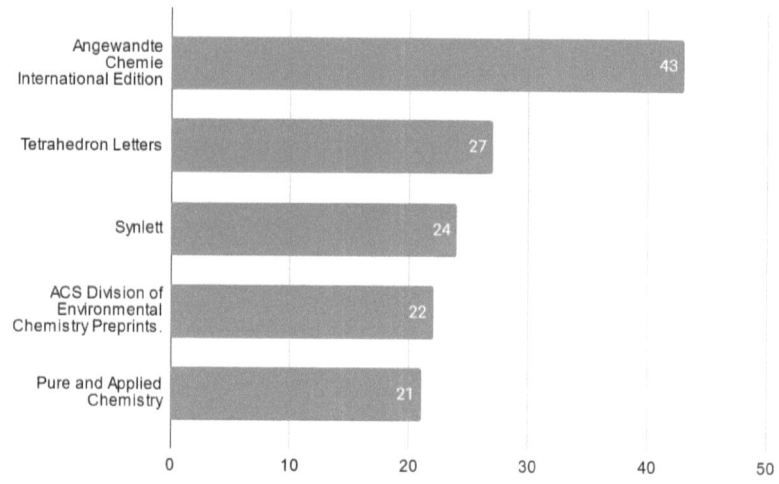

Figure 8: Five major journals/series publishing green chemistry articles (excluding *Green Chemistry*) between 2001 and 2005

The high place of *Angewandte Chemie* when it comes to the number of green chemistry articles coincides with the growing proportion of Germany-based scholars identifying themselves with the movement. These were, however, not the European but the East Asian countries where the interest in green chemistry was growing the most rapidly, with Japan overtaking the UK, and China distancing France. Nevertheless, the primacy of the US remained unchallenged between 2001 and 2005 (Figure 9).

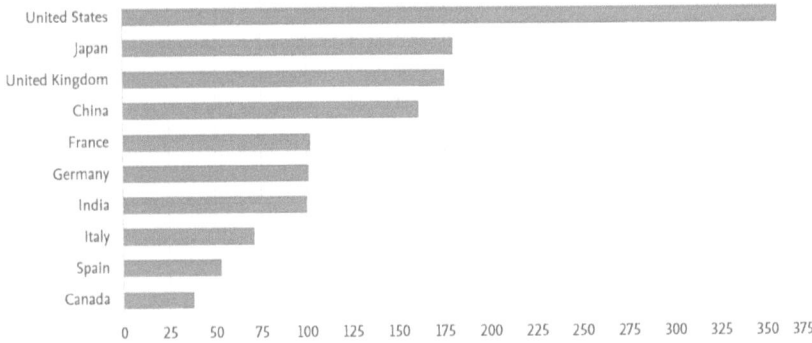

Figure 9: Green chemistry articles by national affiliation of the authors' institutions (2001–2005)

These national shifts did not radically transform however the structure of the leadership in the field. In the period concerned, Clark was definitely the most productive scholar. He contributed to no less than 34 articles. Other important pioneers include pioneers such as M. Poliakoff (UK), R. D. Rogers (US), A. Loupy (France), and J. D. Holbrey (UK). Anastas occupies the thirteenth position with 'only' 9 articles, but he remains among the most cited researchers. His "Origins, current status, and future challenges of green chemistry" is another conceptual and foundations-laying article for which Anastas became famous (Table 21).

Table 21: Most cited green chemistry articles published between 2001–2005 (in bold papers published in *Green Chemistry*; citation count as of 1st of June 2021)

	Title	Year	Citation count
1	Characterization and comparison of hydrophilic and hydrophobic room temperature ionic liquids incorporating the imidazolium cation[10]	2001	3266

---

10 Jonathan G. Huddleston, Ann E. Visser, W. Matthew Reichert, Heather D. Willauer, Grant A. Broker, and Robin D. Rogers, "Characterization and comparison of hydrophilic and hydrophobic room temperature ionic liquids incorporating the imidazolium cation," *Green Chemistry* 3 (2001): 156–164.

2	Origins, current status, and future challenges of green chemistry[11]	2002	1577
3	Sustainable Bio-Composites from renewable resources: Opportunities and challenges in the green materials world[12]	2002	1546
4	**A short history of ionic liquids – From molten salts to neoteric solvents**[13]	2002	1341
5	**Green solvents for sustainable organic synthesis: State of the art**[14]	2005	1204

Table 3 clearly shows that 2001–2005 was the era of solvents, and in particular ionic liquids, with three out of the five most cited articles exploring the topic. It should be noted that article 1 got almost half of the citations the foundational 1998 book got. Three out of these articles were published in *Green Chemistry*, confirming the journal's leading role in the development of the discipline. Does the analysis of keywords indicate any interesting tendencies (Figure 10)?

---

11 Paul Anastas and Mary M. Kirchhoff, "Origins, Current Status, and Future Challenges of Green Chemistry," *Accounts of Chemical Research* 35 (2002): 686–694.
12 Ajit K. Mohanty, Manjusri Misra & Lawrence T. Drzal, "Sustainable Bio-Composites from renewable resources: Opportunities and challenges in the green materials world," *Journal of Polymers and the Environment* 10 (2002): 19–26.
13 John Wilkes, "A short history of ionic liquids – From molten salts to neoteric solvents," *Green Chemistry* 4 (2002): 73–80.
14 Roger A. Sheldon, "Green solvents for sustainable organic synthesis: State of the art," *Green Chemistry* 7 (2005): 267–278.

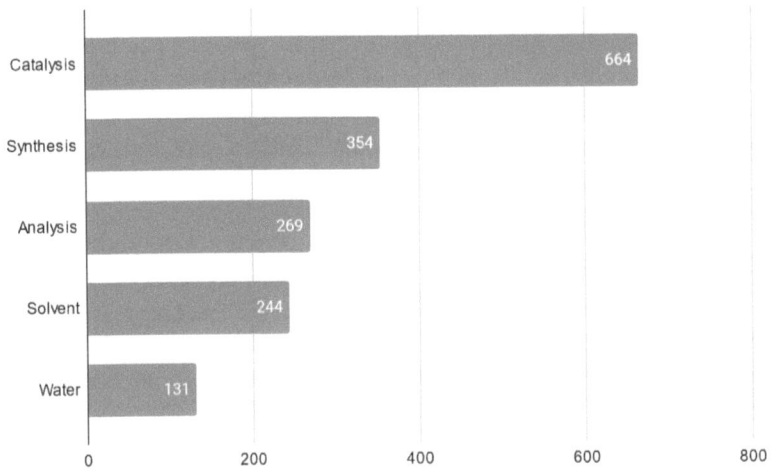

Figure 10: Most popular keywords in green chemistry papers (2001–2005)

According to figure 10, solvents occupy the honourable fourth mention, but the vast majority of articles refer above all to catalysis and synthesis. The last of the most popular keywords is water, suggesting the use of water as a solvent, or the reactions with water as waste. In general, however, one thing leaves no doubt. Once the green chemistry reached the wider community of chemists, it developed, as I have already explained, as the type of chemical synthesis. The most popular keywords do not hint at any movement towards toxicology and broader environmental issues, which was still somewhat the case between 1996 and 2000.

4. New publication venues (2006–2010)

If the years 2001–2005 were a period of stable and continuous growth of the field and its stabilization around broad themes, the following years were somewhat more turbulent. First of all, the number of green chemistry articles published yearly grew by 250% between 2006 and 2010. In total, 3729 articles were published over the period, but only 1221 came from *Green Chemistry*. The journal was then responsible for only a third of the publications, whereas for the years 2001–2005 it had been around 50% (again, these numbers are biased, in the sense that I include all papers published in *Green Chemistry* in the analysis). This is partly due to the establishment of some new important venues. Firstly, in 2007, was created *Green Chemistry Letters and Reviews*. While *Green*

Chemistry published articles from scholars all around the world, it was originally a UK-based endeavour. Anastas and his entourage were rare guests in its pages privileging other ways of communication. Meanwhile, *Green Chemistry Letters and Reviews* was an American initiative with John Warner, the co-author of the seminal 1998 book, being its first editor, and Anastas publishing in the journal's first issue. The introductory editorial fully recognised the value of the *Green Chemistry* journal with which the new venue was not meant to compete. Its goal was to focus on the rapid communication of results. One of the journal's selling points was the use of internet-based tools to provide, for example, "board/chat rooms … to allow direct and public correspondence with authors regarding their published work."[15] The *Letters* continued to play an important role in the development of the reflection of green chemistry, in particular in the field of green chemistry education.

Another important newcomer in the field was *ChemSusChem* created in 2008. The keyword of the new publication was, and still is, sustainability. The journal's first editorial states that

> ChemSusChem covers research on sustainable (green) chemistry, catalysis and biocatalysis, renewable energies and resources, biomass and biofuels, solar energy conversion and photovoltaics, hydrogen storage and fuel cells, carbon dioxide capture and storage, biodegradable products and chemical recycling, as well as sustainability-related aspects of analytical, environmental, and cultural heritage chemistry, and much more.[16]

As we can see, the editorial conflates green and sustainable chemistry, but it enumerates it among other topics the journal was supposed to study. As a consequence, I do not treat all *ChemSusChem*'s papers as, by definition, green chemistry articles. Unlike in the case of *Green Chemistry* and *Green Chemistry Letters and Reviews*, I do not incorporate all articles published in its pages by default into the analysis. Another argument for this choice is that its authors very often tagged their articles with the "green chemistry" keyword, or mentioned it in the abstracts, in order to underline the specificity of their approach as compared to other papers published in the journal.

The third important publication that began in 2009 was the Royal Society of Chemistry's book series *Green Chemistry*. Between 2009 and 2020 more than 70 volumes have been published. These books had usually (but not always) multiple authors who contributed with their studies on specific topics of choice. Scopus usually counts every chapter as a single article. While the series is published by

---

15 John Warner, "The natural evolution of green chemistry," *Green Chemistry Letters and Reviews* 1 (2007): 1–2.
16 Anjum Dadabhoy and Peter Gölitz, "Sustainability: Chemistry Is Key," *ChemSusChem* 1 (2008): 4–5.

the Royal Society of Chemistry, like the *Green Chemistry* journal, its scope is much larger, often including topics on sustainability in general. For example, there is a volume entitled *Sustainable Solutions for Modern Economies* with a chapter on sustainable finance.[17] Sustainable finance is of course outside the scope of green chemistry properly speaking and at least some volumes barely refer to green chemistry concepts at all. Still, I incorporate them in the scientometric analysis due to the fact that the series belongs specifically to the same tradition as the *Green Chemistry* journal and makes part of the canon of the discipline (and has green chemistry in its name!). It is important to note that the opening of green chemistry towards sustainability in general, while present in green chemistry publications from the beginning, became increasingly pronounced starting from the late 2000s. *ChemSusChem* and the eclectic approach of the *Green Chemistry* series are some indicators of this trend.

When it comes to the overall share of the green chemistry articles published between 2006 and 2010, *Letters and Reviews* count 120 papers (all of them included in the count) and they were followed by *ChemSusChem* and then by more general chemistry publications, such as the French *Info Chimie Magazine* targeting a wider public (Figure 11).

---

17 Rainer Hofer, ed., *Sustainable Solutions for Modern Economies* (Cambridge: RCS Publishing, 2009).

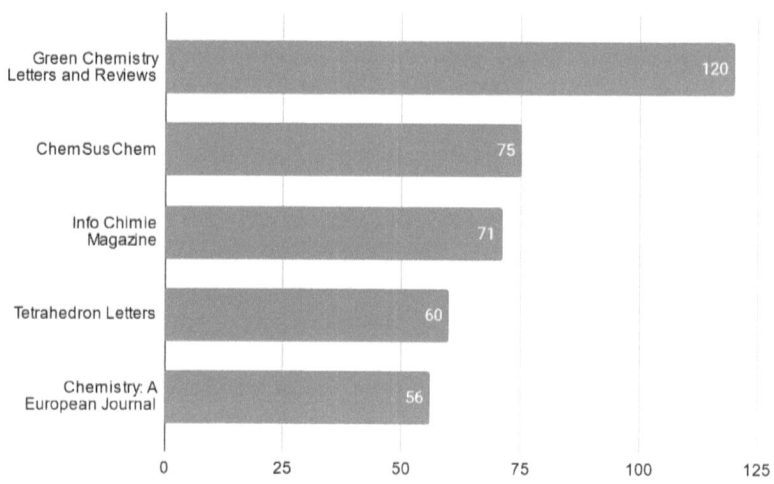

Figure 11: 5 major journals/series publishing green chemistry articles (excluding *Green Chemistry*) between 2006 and 2010

Another reason why the period was special was the fact that the number of articles from India and China was exploding overall and threatened the US primacy in the discipline (Figure 12).

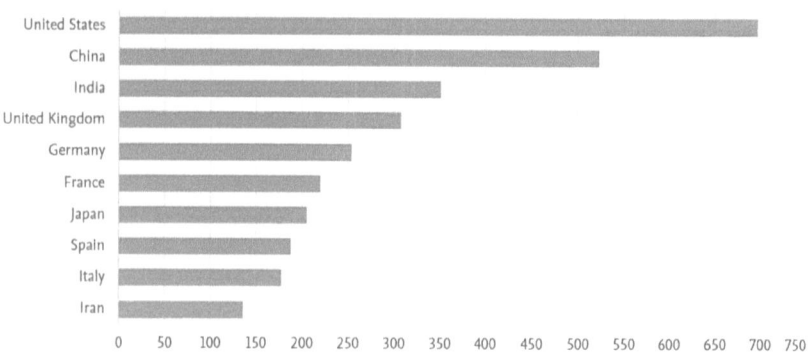

Figure 12: Green chemistry articles by national affiliation of the authors' institution (2006–2010)

Not only was the number of articles and publication venues growing but what also changed were the topics labelled as green chemistry by the researchers in the field. In addition to synthesis and catalysis, there were two important newcomers in the list of the top 5 keyword categories: nanoparticles and plant-related (Figure 13). The ionic liquids clearly started falling from grace.

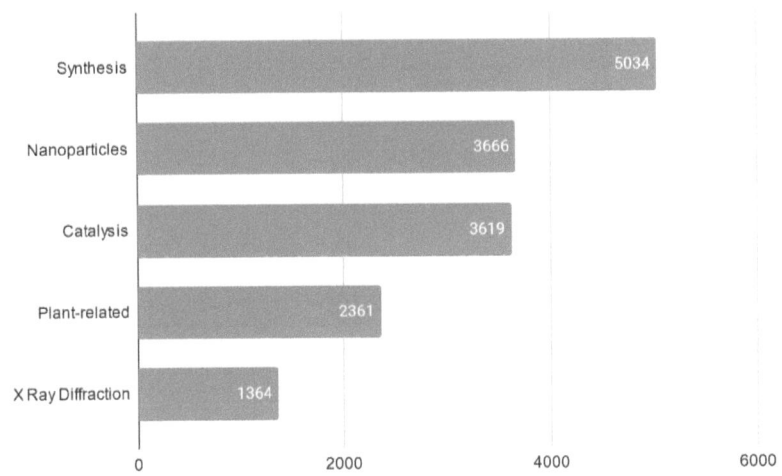

Figure 13: Most popular keywords in green chemistry papers (2006–2010)

What is the relationship between nanoparticles, plants, and green chemistry? The idea is to use plants (their leaves, roots, stems, etc.), or fungi to synthesise nanoparticles. As explains one summary of this line of research: "Greener synthesis of nanoparticles provides advancement over other methods as it is simple, cost-effective, and relatively reproducible and often results in more stable materials."[18] Synthesis of molecules using plants in ambient temperature and without dangerous solvents and non-recoverable catalysts is probably as close to the ideal green chemistry as we can reach. Nevertheless, it is necessary to explain that these studies stem to a large extent from the (white) biotechnology community, not the green chemistry one. The biotechnology scholars appropriated the fashionable term developed by green chemists to describe their scientific work. This appropriation is by no means abusive, the biosynthesis of nanoparticles appears to satisfy the criteria for greenness traditionally construed. At the same time, these scholars rarely refer to the 12 principles framework, or to green chemistry

---

18 Oxana V. Kharissova, H. V. Rasika Dias, Boris I. Kharisov, Betsabee Olvera Pérez, Victor M. Jiménez Pérez, "The greener synthesis of nanoparticles," *Trends in Biotechnology* 31 (2013): 240–248.

metrics, or to the LCA for that matter, which of course poses the risk that some of these bio-syntheses may be demonstrated to be less green in the future. What matters as of now is that in the period 2006–2010, the popularity of the term green chemistry grew rapidly, but it came to incorporate concepts and ideas more and more beyond its initial scope. This shift is also illustrated by the most cited papers published between 2006 and 2010 (Table 22).

Table 22: Most cited green chemistry articles published between 2006–2010 (in bold papers published in *Green Chemistry*; citation count as of 1st of June 2021)

	Title	Year	Citations
1	A safe operating space for humanity[19]	2009	5185
2	**Technology development for the production of biobased products from biorefinery carbohydrates[20]**	2010	2532
3	A green approach to the synthesis of graphene nanosheets[21]	2009	1849
4	Green Chemistry: Principles and Practice[22]	2010	1848
5	**Catalytic conversion of biomass to biofuels[23]**	2010	1544

Table 22 is important for a wide range of reasons. First of all, for the first time, the majority of the top 5 does not come from *Green Chemistry*, a tendency that will become increasingly pronounced in the following years. Secondly, there is another overview article by Anastas, "Green Chemistry: Principles and Practice" published in *Chemical Society Reviews*. This becomes a pattern at this stage. Anastas published throughout his career numerous articles with similar titles restating original principles and presenting the state of the art as well as new promising lines of research.

---

19 Johan Rockström, Will Steffen, Kevin Noone, et al. "A safe operating space for humanity," *Nature* 461 (2009): 472–475.
20 Joseph J. Bozell and Gene R. Petersen, "Technology development for the production of biobased products from biorefinery carbohydrates," *Green Chemistry* 12 (2010): 539–554.
21 Hui-Lin Guo, Xian-Fei Wang, Qing-Yun Qian, Feng-Bin Wang, Xing-Hua Xia, "A Green Approach to the Synthesis of Graphene Nanosheets," *ACS Nano* 3 (2009): 2653–2659.
22 Paul Anastas and Nicolas Eghbali, "Green Chemistry: Principles and Practice," *Chemical Society Reviews* 39 (2010): 301–312.
23 David Martin Alonso, Jesse Q. Bond and James A. Dumesic, "Catalytic conversion of biomass to biofuels," *Green Chemistry* 12 (2010): 1493–1513.

These recurrent publications make it possible to keep the concept of green chemistry, as imagined by Anastas, relevant to subsequent generations of scholars. If someone enters the field, there is always a recent article by Anastas outlining the key definitions and hot topics at any given moment. These articles are useful introductions for chemists and a precious tool for historians, but they can also be seen as a form of self-promotion, and sometimes even of distortion of the historical narrative.

Another intriguing observation concerning the most cited articles from the period concerns the two articles published in *Green Chemistry*. They both contribute to studies on biomass valorisation, so to principle 7 (renewable materials). This is interesting because principle 7 was the least solidly rooted in the general EPA green chemistry philosophy. It was certainly not the core element of what green chemistry originally was in the late 1990s. And yet for Clark, this principle grew in importance over the years. The fact that these popular articles are specifically devoted to this principle marks another shift in the field I discuss in detail in the following chapter.

The last important point of this table is the most-cited article, "A safe operating space for humanity" published in *Nature*, which is, in fact, the most cited article tagged as green chemistry ever according to Scopus. The article explores the problem of "planetary boundaries" that humanity should not exceed in order to preserve the planet. The problem is that the article does not mention green chemistry at any point and cites no green chemistry-relevant sources. It was tagged with the "green chemistry" keyword in the Scopus database (as well as in PubMed), either by the authors or by the data providers. If the article can be considered green chemistry, it is the greenness in the early 1990s-sense, closely associated with environmental sciences broadly construed, and not with what green chemistry was supposed to be according to the EPA and Anastas. Whatever the intentions were, this publication reveals, again, problems with the way the term green chemistry is understood and used.

5. Rise of Asia (2011–2015)

The following five years (2011–2015) were certainly 'calmer' than the previous ones, with some shifts and changes, but without any major revolution. The relative growth of green chemistry articles slowed down, even though it was still impressive in absolute terms. The most remarkable change in these 5 years was the shift on the podium of the most productive nations. China and India clearly dominated the field and outpaced, for the first time, the United States. At the same time, all the three leading nations distanced all their competitors (Figure 14).

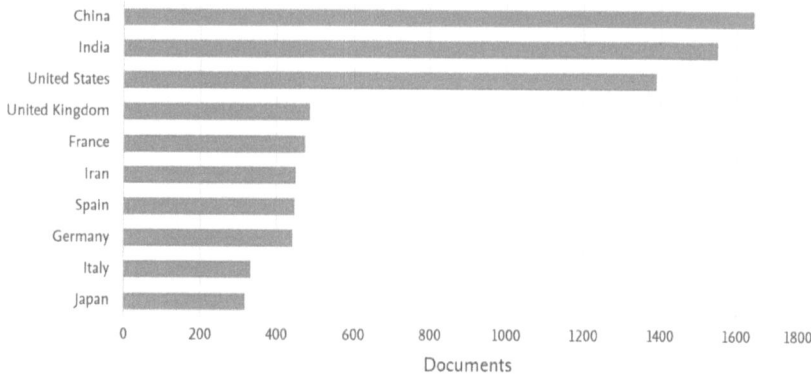

Figure 14: Green chemistry articles by national affiliation of the authors' institution (2011–2015)

As for the keywords, no major change shift is observable, with synthesis, catalysis, and nanoparticles reaffirming their position and distancing all the others (Figure 15). The importance of water (in the context of solvents) and analytical tools such as X-ray diffraction is also recognizable. Again, it is the list of the most cited articles that reveals more interesting tendencies (Table 23).

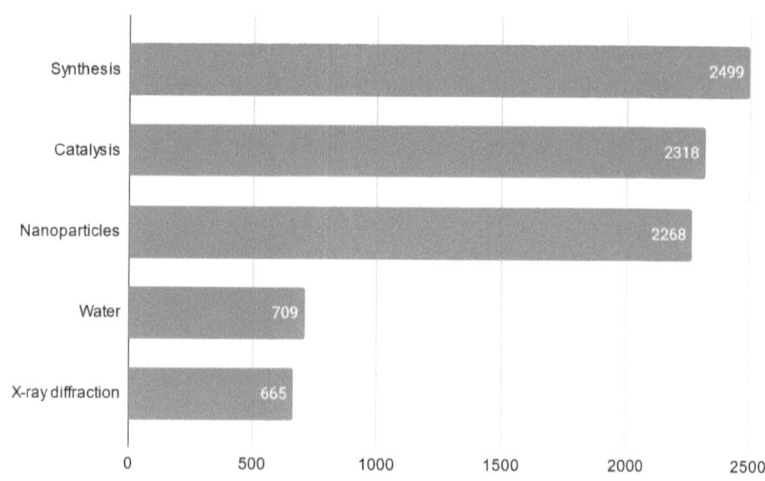

Figure 15: Most popular keywords in green chemistry papers (2011–2015)

Table 23: Most cited green chemistry articles published between 2011–2015 (in bold papers published in *Green Chemistry*; citation count as of 1st of June 2021)

	Title	Year	Citations
1	Deep eutectic solvents: Syntheses, properties and applications[24]	2012	2079
2	Conversion of biomass to selected chemical products[25]	2012	1616
3	**Green synthesis of metal nanoparticles using plants[26]**	2011	1514
4	Engineering the third wave of biocatalysis[27]	2012	1473
5	The cross-dehydrogenative coupling of Csp3 -H bonds: A versatile strategy for C-C bond formations[28]	2014	1232

An article on deep eutectic solvents, a new promising category of ionic liquids, was the most cited one from the period. It was followed by an article on biomass, and then by a paper on the plant-based synthesis of nanoparticles. Remarkably, only one of the top 5 articles comes from *Green Chemistry*, which is correlated with the fall of the overall share of the journal in green chemistry publications. Out of 8972 articles tagged as green chemistry in Scopus between 2011 and 2015, 2439 came from *Green Chemistry*, substantially less than one-third.

Interestingly, in the top 5 journals publishing green chemistry articles other than *Green Chemistry*, there is the *Journal of Chemical Education*, so not a 'scientific' journal strictly speaking but a one focused on pedagogy and communication. Green chemistry was a topic more and more explored in the context of knowledge transmission. Otherwise, the first two positions are occupied by the

---

24 Qinghua Zhang, Karine De Oliveira Vigier, Sébastien Royer and François Jérôme, "Deep eutectic solvents: syntheses, properties and applications," *Chemical Society Reviews* 41 (2012): 7108–7146.
25 Pierre Gazellot, "Conversion of biomass to selected chemical products," *Chemical Society Reviews* 41 (2012): 1538–1558.
26 Siavash Iravani, "Green synthesis of metal nanoparticles using plants," *Green Chemistry* 13 (2011): 2638–2650.
27 Uwe Bornscheuer, Gjalt Huisman, Romas Kazlauskas et al. "Engineering the third wave of biocatalysis," *Nature* 485 (2012): 185–194.
28 Simon A. Girard, Thomas Knauber, Chao-Jun Li, "The cross-dehydrogenative coupling of C(sp3)-H bonds: a versatile strategy for C-C bond formations," *Angewandte Chemie – International Edition* 53 (2014): 74–100.

RCS book series and by the *Letters*. In both cases, all the papers published in both venues are included in the figure (Figure 16).

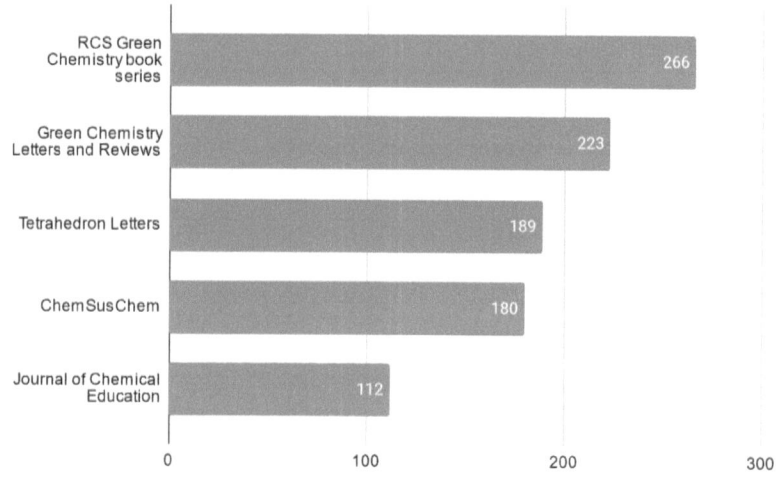

Figure 16: Five major journals/series publishing green chemistry articles (excluding *Green Chemistry*) between 2011 and 2015

6. Solidifying change (2015–2020)

The most recent 5-year period is the one of a steady continuing growth. The number of green chemistry articles published yearly reached 3500 in 2020 and nothing indicates that the trend is about to be reversed. Among 14 779 articles published in total between 2016 and 2020, only 3300 came from the *Green Chemistry* journal, less than one-fourth, meaning its overall share continued to dwindle as well as its importance in defining what green chemistry should be (Figure 17).

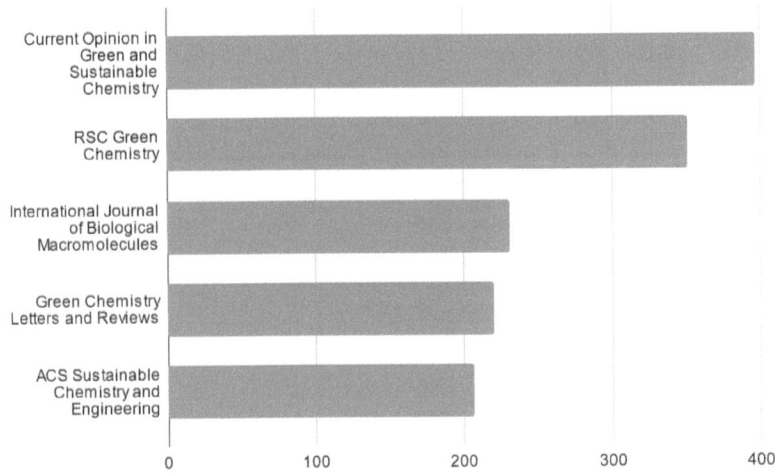

Figure 17: Five major journals/series publishing green chemistry articles (excluding *Green Chemistry*) between 2016 and 2020

In figure 17, there are three interesting newcomers. First of all, the period was marked by the creation of a new important journal: *Current Opinion in Green and Sustainable Chemistry*. Klaus Kümmerer, the editor of the journal, did not position his publication towards other reviews in any manner. It seems, however, that the purpose of the review was to integrate the terms green chemistry and sustainable chemistry. Kümmerer, faithful to the German understanding of the field, argues that sustainable chemistry is a more all-embracing concept than green chemistry, a topic to which I return later.

Another new venue on the list is the *International Journal of Biological Macromolecules*. Its high position results from the further development of topics such as biosynthesis of nanoparticles as well as the emergence of new bridges between chemistry and molecular biology in general. Finally, the *ACS Sustainable Chemistry and Engineering* was a review created in 2013 that came to play an important in uniting green and sustainability frameworks, although with a less clearly established philosophy than *Current Opinion in Green and Sustainable Chemistry*.

When it comes to productivity on the international level, previously visible trends became increasingly accentuated without major changes. Chinese scientists published twice as many articles in green chemistry as their American counterparts and India gained a safe second place (and Iran the fourth one). Asia fully embraced the green chemistry nomenclature meaning that the frontiers of the field will be decided more and more outside the Western scientific communities that had initially invented the term (Figure 18). It does not mean the traditional centres of green chemistry said their last word though. In spite of this new international shift of power relations, between 2016 and 2020 James Clark was again the most productive researcher in green chemistry, 20 years after the creation of his journal.

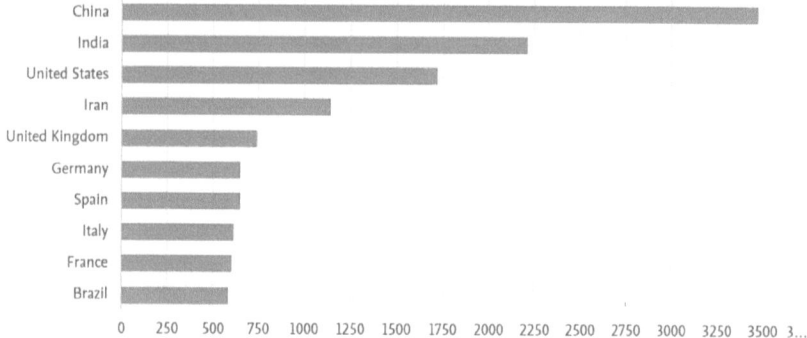

Figure 18: Green chemistry articles by national affiliation of the authors' institution (2016–2020)

As for the keywords (Figure 19), they reveal some noticeable changes. Most interestingly the keyword category "nanoparticles" comes triumphant for the first time, outclassing both synthesis and catalysis. The fourth place is occupied anew by plant-related topics (excluding, however, the term biomass).

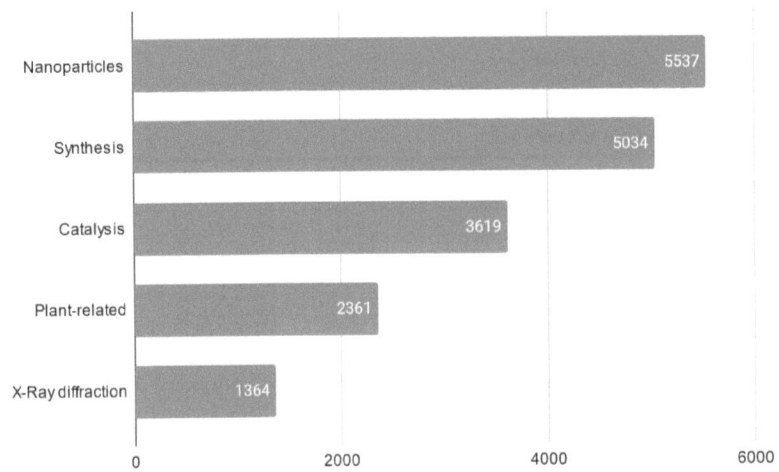

Figure 19: Most popular keywords in green chemistry papers (2016–2020)

When it comes to the most cited articles (Table 24), without forgetting that the exercise is delicate in such a short time span, it mostly confirms the keyword analysis, with the most popular article being on silver nanoparticles. Interestingly, the second most cited article features, for the first time, the life cycle analysis; a tendency I pointed out in the previous chapter with the LCA taking more and more importance among people identifying with green chemistry. Other than that, only one of these articles comes from the *Green Chemistry* journal, suggesting again that it may be losing its grip over the discipline. Overall, biocatalysis and biosynthesis, appear to be some of the hottest topics in the field of green chemistry in the second half of the 2010s.

Table 24: Most cited green chemistry articles published between 2016–2020 (in bold papers published in *Green Chemistry*; citation count as of 1st of June 2021)

	Title	Year	Citations
1	Silver nanoparticles: Synthesis, characterization, properties, applications, and therapeutic approaches[29]	2016	738

---

29 Xi-Feng Zhang, Zhi-Guo Liu, Wei Shen, Sangiliyandi Gurunathan, "Silver

2	Sustainable Conversion of Carbon Dioxide: An Integrated Review of Catalysis and Life Cycle Assessment[30]	2017	613
3	Wood-Derived Materials for Green Electronics, Biological Devices, and Energy Applications[31]	2016	591
4	**Efficient, selective and sustainable catalysis of carbon dioxide**[32]	2017	492
5	Role of Biocatalysis in Sustainable Chemistry[33]	2017	459

7. Concluding remarks

The vast set of data presented in this chapter reveals a range of very interesting phenomena, especially if combined with a slightly more thorough reading of some of the most cited articles in question. First of all, the different periods I analysed are not really comparable from the thematic standpoint. The number of articles concerning ionic liquids published in the early 2000s is not comparable to the sheer amount of works on the biosynthesis of nanoparticles and biocatalysis in the late 2010s. Certainly, we can say that fashions changed, but if one were to analyse the entire corpus of literature in green chemistry as if it was a single coherent field, some topics that captured the imagination of chemists in the late 1990s or the early 2000s would become overall a minor footnote in the discipline's profile.

Another obvious tendency is that the number of articles labelled as green chemistry grew exponentially every few years between 1996 to 2020 and the end of this trend is nowhere to be seen. At the same time, the overall share of the *Green Chemistry* journal has been decreasing. It corresponded to half of all the

---

Nanoparticles: Synthesis, Characterization, Properties, Applications, and Therapeutic Approaches," *International Journal of Molecular Sciences* 17 (2016): 1534.
30 Jens Artz, Thomas E. Müller, Katharina Thenert et al., "Sustainable Conversion of Carbon Dioxide: An Integrated Review of Catalysis and Life Cycle Assessment," *Chemical Reviews* 118 (2018): 434–504.
31 Hongli Zhu, Wei Luo, Peter N. Ciesielski et al., "Wood-Derived Materials for Green Electronics, Biological Devices, and Energy Applications," *Chemical Reviews* 116 (2016): 9305–9374.
32 Qing-Wen Song, Zhi-Hua Zhoua and Liang-Nian He, "Efficient, selective and sustainable catalysis of carbon dioxide," *Green Chemistry* 19 (2017): 3707–3728.
33 Roger A. Sheldon and John M. Woodley, "Role of Biocatalysis in Sustainable Chemistry," *Chemical Reviews* 118 (2018): 801–838.

articles in the early 2000s, but fell to less than 25% by the late 2010s. This is also due to the appearance of numerous new publication venues explicitly referring to green chemistry. The impact of this shift is not to be underestimated because these other reviews and journals often connect to different traditions or disciplinary frameworks than Clark's *Green Chemistry*, and their editorial boards may use a more intuitive definition of greenness. This applies, to an even larger degree, to the general chemistry reviews in which green papers are published. As explained in the previous chapter, even the editorial board of the flagship journal such as *Green Chemistry* struggled to define admission criteria, but at least the discussion over this difficult topic took place and the key scientists fully recognized that there was a challenge on this front. The problem is that when we read some of the most cited articles in the field published outside the pages of *Green Chemistry*, they often do not mention anything like the 12 principles or any other criteria for evaluating greenness for that matter.

This is, for example, the case of the most cited article from the period 2016–2020 on the biosynthesis of silver nanoparticles.[34] The supposed greenness of the method in question results exclusively from the use of plants in the process. I do not deny the potential eco-friendliness of the article's chemistry, which is perhaps of great quality. It is not, however, the green chemistry as defined by the theoretical developments in the field. Many of these most cited articles do not demonstrate their greenness at all, and it is usually implied because of the use of plants or biomass. However, as much as people assumed that every ionic liquid is a green solvent, at some point one may wonder whether every biosynthesis or every biocatalyst is also green. The key learning of this section is that in spite of the fact that green chemistry's theoretical foundations greatly developed over the last two decades through new sets of principles and lively debates on what is green and what is not, there is little evidence that the wider community of scientists practising what they call green chemistry are familiar with them at all. Or to put it simply: a tremendous amount of green chemistry research remains very intuitive, in spite of the fact that we know now better than ever this 'gut feeling' green chemistry is full of pitfalls.

---

34 Zhang, Liu, Shen, and Gurunathan, "Silver Nanoparticles: Synthesis, Characterization, Properties, Applications, and Therapeutic Approaches."

8. Side note on patents.

Before passing on to the next chapter, I have to address the elephant in the room: why have I not included the patenting trends in the graphs above? After all, green chemistry is all about practical solutions that are supposed to help the chemical industry to become more environmentally friendly. As a consequence, it may seem that a scientometric study of green chemistry patents should be the most obvious way of evaluating the field's real contributions. Two major studies on green chemistry patents conducted in recent years deserve particular attention. One from 2004, entitled "Adoption of Green Chemistry: an analysis based on U.S. patents"[35] and another one from 2020 published under the title: "What Do Patents Tell Us About the Implementation of Green and Sustainable Chemistry?."[36] Both are comprehensive works, rich in details, and worth reading to better understand the dynamics of the green patenting in chemistry. Both of them, I argue, fail to recognize some fundamental flaws in their methodology. In fact, in order to make any meaningful query in a patent database, one must choose relevant keywords. But the choice of keywords itself is a normative act as it draws the perimeter of what authors think that green chemistry is. Both articles study very well what the authors believe to be green chemistry but without accounting for the problems with the concept itself.

Firstly, I have a look at the 2004 publication. The authors explain that the article's

> iterative search strategy [was] built on the definition and keywords describing green chemistry in Anastas and Warner (1998) and areas of green chemistry identified by the OECD Sustainable Chemistry initiative (Tundo et al., 2000) [co-authored by Anastas] and guidance from the EPA's Green Chemistry program ... Search terms were intended to capture concepts such as the use of alternative feedstocks, solvent-free processes, use of innocuous reagents, natural processes, use of alternative solvents, design of safer chemicals, development of alternative reaction conditions, and minimal energy consumption.

The paper then adds: "some patent classifications were included in the filter either to limit the keyword search to chemically related areas of technology or to exclude subject matter related to pollution remediation and waste treatment."

---

35 Tamara J. Namaroff, R. J. Garant, M. B. Albert, "Adoption of Green Chemistry: an analysis based on U.S. patents," *Research Policy* 33 (2004): 959–974.
36 David J. C. Constable, "What Do Patents Tell Us about the Implementation of Green and Sustainable Chemistry?," *ACS Sustainable Chemistry and Engineering* 8 (2020): 14657–14667.

The search is explicitly rooted in the EPA's definition and green chemistry is understood as a category of chemical syntheses satisfying a certain number of criteria. Some of the terms used by the authors are however not very specific, i.e. natural processes, alternative feedstocks, and alternative solvents. What were the exact keywords used for the patent analysis? Unfortunately, we do not know. The exact search terms is not included in the article. Apparently, they were available on the Green Chemistry Institute's website back in 2004 but are not any more. As a consequence, all the detailed results that follow in the article describing the evolution of the patent trends, their division between countries and industrial sectors, as well as the hierarchy of companies in terms of green patent numbers, are based on unknown search terms.

This is only a surface-level problem though. After all, there is some indication about the type of greenness markers that had been searched for. It is the plot the authors obtained concerning the total number of patents filed between 1983 and 2001 that really deserves attention. Their result was the following:

> The number of granted patents was fairly constant from 1983 to 1988, averaging 71 patents per year. Between 1989 and 1994, the number of green chemistry patents granted per year increased rapidly, from 92 to 251. The annual patent count continued to grow after 1994, but at a rate that was closer to the pre-1989 rate than the 1989-1994 rate. From 1995 to 2001, an average of 267 green chemistry patents were granted each year.[37]

In other words, the authors identified three periods: 1983-1988, 1989-1994, and 1995-2001, characterized by different dynamics, the middle one being the period of rapid growth and the other two were marked by a very slow incline. Of course, the entire analysis begs immediately a few questions. First of all, if the purpose of the query was to find green chemistry patents, then we learn that green chemistry was around already back in 1983 with around 71 patents filed yearly. It would be interesting to go back in time even further. Were there any patents satisfying these criteria already back in the 1960s or the 1950s? And if yes, what does it tell us about the supposed novelty of the green chemistry approach, and the green/brown division?

Nonetheless, these results may appear to confirm our intuitions about the field: after the careless 1980s, in the early 1990s the topic of pollution prevention exploded and the industry began treating green chemistry more seriously. But then again, this was not the EPA's green chemistry; the influence of the EPA's

---

37 Namaroff, Garant, Albert, "Adoption of Green Chemistry: an analysis based on U.S. patents."

conceptualisations on "green chemistry" or "benign by design" was minuscule in the early 1990s and, obviously, the EPA's influence cannot explain this major shift in patenting trends that we witness over the period. This was a shift coming from the industry itself.

At the same time, it is worth noting that the number of green patents filed yearly between 1993 and 2001 grew only by 20% in spite of the fact that the total number of US patents filed over the same period grew almost by 75%.[38] This means that the overall share of supposedly green patents in the total pool of patents filed in the US between 1993 and 2001 actually fell. Or to put it differently: despite the fact that green chemistry became formalized (conferences, manuals, journals, institutions) in the 1990s, and the Presidential Green Chemistry Award has been attributed since 1996, there were proportionally fewer green patents in the early 2000s than in the early 1990s.

This result of the 2004 article, and especially the sudden growth of green patenting in the early 1990s, have two possible interpretations. The more charitable one is that the 1990 Pollution Prevention Act was successful in changing the industry's priorities. If this is the case, however, the 1998 book by Anastas and Warner and the entire EPA's green chemistry was a mere codification of what had been already practised by the industry before (I remind that the 2004 article was based on the definition of green chemistry formulated by the EPA). If so, the green chemistry philosophy had actually very little to offer to the industry already implementing what the EPA was preaching. This interpretation is fully in line with the recollections of some of the former industrial chemists who did not see in green chemistry anything particularly revelatory.[39] To be fair, one thing that green chemistry had to offer to the industry was its name. A range of practices that had been previously implemented in the industry due to environmental policies or simple economic calculations could be, from then on, described as "green chemistry," an eco-friendly framework, the plausibility of which was reinforced by the noble genealogy stretching back to Rachel Carson. In this sense, one can come to a somewhat pessimistic conclusion that the green chemistry philosophy was a tool for the industry to hijack the environmental tradition.

---

38 U.S. Patent Statistics Chart Calendar Years 1963–2020, accessed 21/02/2022, https://www.uspto.gov/web/offices/ac/ido/oeip/taf/us_stat.htm.

39 Mark A. Murphy, "Early Industrial Roots of Green Chemistry and the history of the BHC Ibuprofen process invention and its Quality connection," *Foundations of Chemistry* 20 (2018): 121–165.

This was the more charitable interpretation. The less charitable interpretation is that the terms used for the search in the patent database had to inevitably give such a result. These elements were based on the input of Anastas and his colleagues, many of whom started their careers in the early 1990s. They saw in that period studies on a range of emerging innovations they considered to be disruptive and based their view of greenness on these technologies. Their observations translated into the conceptual framework of green chemistry on which the scholars analysing patent trends based their query. The query that led to the plot showing… what technologies were fashionable and promising in the early 1990s. Or to put it differently: the input conditioned the shape of the curve. If so, the entire study was based on the wrong premises.

It may be objected that I am reading too much into this plot and that the 2004 study aimed to simply study the evolution of environmentally friendly patenting over the previous decades. But this is simply not the case, considering that the authors explicitly limited themselves to green chemistry in a narrow sense and, for example, excluded from their analysis pollution remediation or water treatment, which are arguably equally important from the sustainability perspective.

Many of the pitfalls enumerated above were avoided by the second major patent analysis published in 2020. The article was written by one of the most prominent scholars of green chemistry, David J. C. Constable. It was a comprehensive overview of some major patenting trends throughout the history of the green chemistry concept. In his article, Constable notably provides exact queries in patent databases facilitating reproducibility of his results and he engages in the discussion on the pertinence of search keywords and other methodological challenges, making the article one of the most precious sources for understanding the development of green chemistry ever conducted.

His conclusions are pretty pessimistic though. He notes that over the last 30 years only around 1.1% of all patents could be qualified as green chemistry.

> [T]he results of this study suggest that few of the chemistry innovations in academia have found their way into routine commercial applications. An equally disturbing conclusion is that many patents are tied to only one or two green attributes as in the case of pursuing biodegradability in polymers while using monomers that are known to be highly hazardous. Or, pursuing catalytic routes using precious metals whose life cycle environmental, safety, health and sustainability impacts far outweigh the benefits of their use. This is the true nature of chemistry as it is currently practiced, and these

results suggest there remains considerable work to implement greener, more sustainable approaches to making chemicals and all the products that come from them.[40]

Constable offers some explanations as well:

> Industry over the last 30 years has seen a decimation of industrial R&D in the chemical and allied industries that has favored incremental improvements to existing product lines in favor of truly novel, let alone green, research. One need look no further than the relentless pursuit of shareholder value added that has led to the merger, acquisition, demergers, break-ups, and remaking of the chemical industry and the near death of the kind of research that characterized industry prior to 1990, explaining the dependence upon, and promotion of, unsustainable practices.

However, Constable's results can give an even bleaker image than he realizes. To crudely simplify, his queries include terms such as "environmentally benign chemistry" or "bio-derived polymer," or simply "green chemistry." Constable explains problems concerning cleaning the results he obtained by excluding, for example, the hazardous chemicals that would disqualify the greenness claim. But the point is that at least some of the keywords chosen for Constable's study still remain based on self-evaluation. If the researcher puts some green terminology in his patent, Constable's study may simply count it as green. How meaningful is this evaluation then? What does Constable's study really measure? The number of genuinely environmentally friendly patents, or the number of patents using the green and environmental vocabulary? To reiterate: the greenness claim is a powerful political (and marketing) statement, and the companies have vested interests in underlining greenness whenever possible.

It may be objected that the same difficulties apply to any scientometric study. After all, this is exactly the problem I have identified in the sections concerning publications in scientific reviews. And yet, for an academic paper to be published, some reviewers, editors, or editorial boards, have to approve the greenness claim even if the criteria are extremely shoddy. This is the expectation from the more prestigious journals with "green" or "sustainable" in their titles. As a consequence, we can hopefully assume that at least a certain number of articles we analysed in the previous section satisfy some greenness criteria, even if the bar is set low. There is no such guarantee in patenting. The patent office will not refuse a patent for having used the term green to describe a technology in question. The only way to truly evaluate the impact of green chemistry on the

---

40 Constable, "What Do Patents Tell Us about the Implementation of Green and Sustainable Chemistry?".

industry is to conduct a much more meticulous empirical analysis over the period of many years in collaboration with the industry and with specialists from different academic disciplines. An unlikely perspective in any foreseeable future.

Overall, the two articles presented above, as valuable as they are on their own merit, hint at the much deeper problems concerning the definition of greenness and its true impact on the world of the industry.

Chapter 6: Biomass and doubly green chemistry

One green chemistry principle is not exactly like the others. Principle 7 states: "Whenever it is practical to do so, renewable feedstocks or raw materials are preferable to non-renewable ones." The particularity of this principle is due to three reasons. Firstly, it is relatively autonomous towards the others. Other principles are often interconnected (e.g. waste prevention and the use of catalysis, or laboratory safety and the use of non-toxic solvents). Principle 7 can stand on its own and it does not imply any other green chemistry principle. On the contrary, the use of renewable materials can perfectly lead to the generation of toxic and wasteful by-products. Conversely, the raw materials for the most harmless and atomically efficient reaction can come from non-renewable fossil feedstocks. The second reason why principle 7 is particular is that it is probably the oldest of the 12 principles, in the sense, that it was explicitly formulated in one way or another manner long before the 1990s. The final key observation concerning principle 7 is that it came to play a more and more prominent role in the green chemistry community becoming, for many, the central principle and the ultimate criterion of greenness. In this chapter, I highlight the key features of principle 7 and the place of the discussions on biomass and renewability in the green chemistry community.

1. Prehistory of principle 7[1]

The use of renewable feedstocks is obviously nothing new. Humanity relied on wood, resin, animal fat, and various agricultural products for thousands of years. It was only by the end of the nineteenth century, when a great many industries started exploiting derivatives of coal and petroleum after fossil feedstocks had become the centrepiece of organic chemistry. One of the first big battlefields between renewables and nonrenewables was of course fuel. Already in 1899, Kaiser Wilhelm II "was enraged at the Oil Trust of his country, and offered prizes to his subjects and cash assistance … to adapt [plant-based alcohol] to use in the

---

[1] Fragments of this section have been previously published in: Marcin Krasnodębski, "*From Forest Waste to Fuel Tank: The French Quest for Autarkic Fuel Sources after World War I*," Technology and Culture, 62 (2021): 105–127.

industries."[2] In 1906, frustration with monopolistic practices in the national oil industry led Theodore Roosevelt to state that "[i]t is highly desirable that an element of competition should be introduced … [by] putting alcohol used in the arts and manufacturers upon the [tax] free list."[3] Using ethanol as a fuel was one of the most important investigative pathways in the early twentieth century. The famous Ford Model T was adapted to run on both petrol and alcohol.[4] Ford himself was an unabated supporter of ethanol as fuel: "The fuel of the future is going to come from fruit like that sumach out by the road, or from apples, weeds, sawdust — almost anything … There's enough alcohol in one year's yield of an acre of potatoes to drive the machinery necessary to cultivate the fields for a hundred years" he told New York Times in 1925.[5]

While in the US the use of alcohol as fuel remained a topic of heated debates, in Europe some countries enforced nation-wide policies to guarantee the use of domestic agricultural products. For example, potatoes were a driving force of the alcohol industry in Germany, with more than two-thirds of the industrial alcohol produced in this country in 1912 (2,449,000 hectolitres) coming from this vegetable.[6]

An alternative fuel was wood gas: a mixture of carbon monoxide and hydrogen obtained by destructive wood distillation in gasifiers. It was, alongside many other renewable fuels of agricultural origin, at the heart of the autarkic projects of many western nations before World War 2, including the United States, France, Germany, Italy, Switzerland, Hungary, the Soviet Union, Finland, Japan or Sweden.[7] In 1936, Marshal Philippe Pétain, a World War I hero and the future leader of the puppet Nazi Vichy regime, warned his fellow countrymen: "France has no sufficient guarantees against the mortal risk of finding itself, in case of

---

2   No author, "Launching of a Great Industry: The Making of Cheap Alcohol," *The New York Times*, 25 November, 1906, 3.
3   No author, "Editorial," *Washington Post*, 5 May, 1906, 1.
4   Michael S. Carolan, "Ethanol versus Gasoline: The Contestation and Closure of a Sociotechnical System in the USA," *Social Studies of Science* 39 (2009): 421–448.
5   No author, "Ford Predicts Fuel from Vegetation," *New York Times*, 20 September, 1925, 24.
6   Jacques Chevalier, "Le carburant national en Allemagne," *Bois et Résineux*, 30 April, 1922, 1.
7   Helena Ekerholm, "Cultural Meanings of Wood Gas Automobile Fuel in Sweden, 1930–1945," in *Past and present energy societies: How energy connects politics, technologies and cultures*, ed. Nina Möllers and Karin Zachmann (Bielefeld: Transcript Verlag, 2012), 223–247.

hostilities, without enough fuel ... [A] rational use of the forest gas is considered to make an immediate contribution to national security."[8]

However, the ambitions concerning renewable sources of energy were not exclusively political. In 1919, *Scientific American* published an article under a very telling title, "How Long the Oil Will Last," and gave a rather chilling answer: no more than 15 years.[9] This opinion was reaffirmed by Charles F. Kettering, then the vice president of research for General Motors.[10] This resonated in the global press. For example, the French *Chimie et Industrie* explained in 1925 that oil reserves would run out by 1950 and that alternative fuels were desperately needed.[11] In terms of problem framing, asserting independence from foreign importations and the peak oil justified then the shift towards renewables in the 1920s as much as they do today.

Of course, it should be noted that the debate was not just about renewable fuels. For example, Ford envisaged the use of bio-based materials to manufacture cars in the 1930s. The research on the use of renewable agricultural feedstocks in the chemical industry was called "chemurgy" in the United States. The golden age of chemurgy was in the 1930s and the 1940s, but the topic ceased to generate much attention in the 1950s, among others due to the availability of cheap petroleum-based alternatives. Much of the work on the history of chemurgy was conducted in the 1990s by agricultural historian Mark Finlay who showed its relevance to modern debates on industrial ecology.[12]

If chemurgy is a prehistory of the biomass research, our modern interest in renewability stems directly from the oil crisis in the 1970s. It was over the period when more and more chemists, economists, and activists advocated the use of biomass instead of petroleum and coal as raw materials for manufacturing not only fuels but also other materials. If we look at the evolution of the popularity of the term "biomass" over the last few decades, there is a staggering growth in the 1970s. The term "renewable" followed a similar trajectory but it caught up with biomass only in the 2000s (Figure 20).

---

8   Philippe Pétain, "Carburant national et véhicules à gazogène," *Revue hebdomadaire* 17 (1936): 391–402.
9   No author, untitled, *Scientific American*, 3 May, 1919, 459 and 474.
10 C. F. Kettering, "Studying the knocks," *Scientific American*, 11 October, 1919: 364.
11 A. Travers, "La Crise des Carburants les remèdes proposés," *Bulletin de l'Institut du Pin* 17 (1925): 241–243.
12 Mark R. Finlay, "Old Efforts at New Uses: A Brief History of Chemurgy and the American Search for Biobased Materials," *Journal of Industrial Ecology* 7 (2003): 33–43.

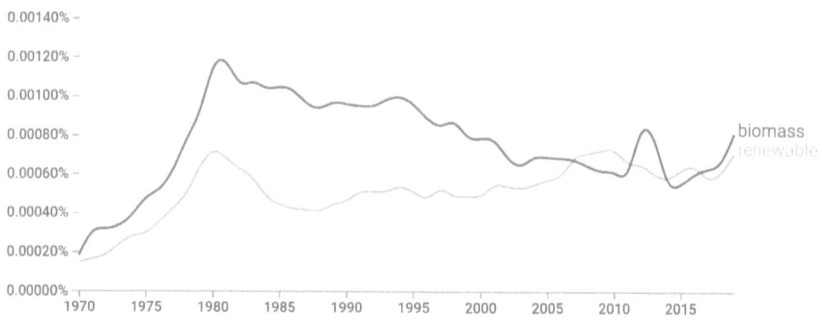

Figure 20: Popularity of terms "biomass" and "renewable" in the English speaking literature between 1950 and 2019 according to Google Ngram (smoothing = 0)

Renewability and biomass valorization were part of the new environmental mindset that crystallized in the 1970s; the same mindset in which green chemistry had its own roots. And yet, as already explained, in the early EPA green chemistry publications from the 1990s, the emphasis was on pollution prevention and not at all on renewability. When Anastas and Warner introduced principle 7 in 1998, it had not had any direct antecedents in the works that made part of this early green chemistry network. The work on biomass and renewable feedstocks was conducted outside of it.

In the countries with different traditions and understandings of green chemistry, the accents were, however, put differently. The first book with green chemistry in its title, *La chimica verde*, published in 1987 in Italy, was precisely about biomass and its uses.[13] It is also in this sense that the term green chemistry was originally used in French publications, where it appeared, as we remember, as a shorthand for plant chemistry in general. Interestingly, in France, this way of understanding green chemistry developed further in competition with the American framing.

2. The French connection

The most comprehensive work on the original character of green chemistry in France was conducted by a French economist Estelle Garnier, who defended her Ph.D. dissertation on the topic in 2012, as well as by her supervisor, Martino

---

13 Italo Pasquon and Luciano Zanderighi, *La chimica verde: le utilizzazioni dei prodotti vegetali e le biotecnologie* (Milano: Hoepli, 1987).

Nieddu, and his team based at the University of Reims Champagne-Ardenne.[14] Just as with Roberts's Ph.D. dissertation, the publications of the French research group are both primary sources about the history of green chemistry, but can also be seen as elements of an evolving narrative shaping the understanding of the concept. Garnier and Nieddu were fully aware of the French exceptionalism or at least of the conflicting vocabularies and made it a topic of numerous studies. Nieddu's later works are particularly valuable, so far as critical epistemology of green chemistry's foundations is concerned, thanks to his close collaboration with French chemists.[15] Nieddu's arguments are not restated here in their entirety, but it is crucial to point out that his work indicated that there was at least a certain degree of controversy among the French scholars concerning the utility of the concept of green chemistry.

To understand the French context, it is necessary to start with the first French textbook entitled *Chimie verte [Green Chemistry]* edited by Paul Colonna, a research director in the INRA (French National Agricultural Research Institute).[16] It was published in 2006, eight years after the formulation of the twelve principles, seven after the establishment of Clark's *Green Chemistry*, and in the middle of a rapid expansion of the field. And yet, Colonna's volume makes almost no reference to the American concept and is entirely devoted to plant and biomass chemistry (*chimie du végétal*). In the book, there are two broad categories of chapters: a few transversal and overarching ones and the majority focused on narrow techno-scientific problems. In the latter category, there are chapters on the lignocellulosic biomass treatment, starch-based products, plant fibres, and biofuels. The transversal articles are mostly devoted to 'big issues' such as the place of agriculture and forestry in the debates on climate change, biodegradability, dangers of aquatic pollution, and economic aspects of non-fossil fuels.

The particularity of the book's message is that its authors perfectly realize that bio-sourced (and renewable) is not equivalent to biodegradable and safe. They underline the necessity of developing non-stoichiometric reactions by using catalysis, using less toxic reactants and solvents, and researching more selective

---

14 Estelle Garnier, "Une approche socio-économique de l'orientation des projets de recherche en chimie doublement verte" (PhD diss., University Reims Champagne-Ardenne, 2012).
15 E.g. Martino Nieddu, Franck-Dominique Vivien, Estelle Garnier, Christophe Bliard, "Existe-t-il réellement un nouveau paradigme de la chimie verte?," *Natures Sciences Sociétés* 22 (2014): 103–113.
16 Paul Colonna, ed., *Chimie Verte* (Paris: Technique Et Documentation, 2006).

processes.[17] As such Colonna's perspective is not necessarily antithetical to the one developed in the EPA. On the contrary, there are many shared elements, such as the focus on reducing toxic wastes. However, Colonna's green chemistry is clearly stemming from a different tradition, ingrained in agricultural, forest, and environmental studies broadly construed developed in the French INRA. Garnier and Nieddu consider Colonna's publication a landmark in the development of what they call "doubly green chemistry" (*chimie doublement verte*), in which pollution prevention is intimately intertwined with the problem of biomass valorization, of bio-sourced materials, and, in particular, of bio-refineries that are heralded as the future of the chemical industry in the post-petroleum world.

Garnier's work explores the social and economic underpinnings of doubly green chemistry all over the world, insisting on the fact that green chemistry is (or rather was in the end of the 2000s when she conducted her research) understood that way in many other countries.[18] The green chemistry networks in Japan, Spain, Canada, and in particular Brazil all used in their mission statements explicit references to the plant and biomass chemistry, well beyond simple reference to principle 7. For some of these networks, green chemistry appeared to be above all an attempt to construct an economy free from fossil-based materials and some of them defined, at least implicitly, the term in the same way as Colonna.

This does not mean this way of thinking of green chemistry was universal in France. Green chemistry understood in the 'American way' had also its proponents. Isabelle Rico-Lattes, an influential French chemist, explained in 2012 that "'green chemistry' does not only refer to the 'chemistry of renewable feedstock,' even if this is frequently how it is perceived, but refers to the 12 principles developed by Paul Anastas."[19] In other words, French chemists outside the INRA's network were not necessarily familiar with the tradition developed in the agricultural chemistry circles. This struggle between the two traditions can be observed in many other French publications over the last fifteen years.

A more Anglo-Saxon understanding of green chemistry can be found for example in the works of Stéphane Sarrade whose books *La chimie d'une planète*

---

17 Colonna, *Chimie Verte*, x.
18 Garnier, "Une approche socio-économique de l'orientation des projets de recherche en chimie doublement verte," 98.
19 Cited in Jean-Pierre Llored and Stephane Sarrade, "Connecting the philosophy of chemistry, green chemistry, and moral philosophy," *Foundations of Chemistry* 18 (2016): 125–152, 141.

durable [*Chemistry for Sustainable Planet*] published in 2011[20] and *Ressources de la Chimie Verte* [*Green Chemistry Resources*] from 2008[21] discuss at length both the principles of green chemistry and of green engineering, even if he also tries to ground green chemistry in broader considerations concerning the renewability of feedstocks and sustainability in general. In 2011, Christine Ducamp, a chemistry professor from Toulouse, positioned herself in the American trajectory and described the 12 principles as constitutive for the concept of green chemistry, but she did it in a publication on agricultural chemistry education.[22] A year later, two French engineers wrote an article equating green chemistry with bio-sourced chemistry without mentioning Anastas or the 12 principles at all.[23] The most openly 'pro-American' French manual that fully embraced the EPA tradition was published in 2017. In the introductory chapter of a comprehensive overview *Chimie verte: Concepts et applications* [*Green Chemistry: Concepts and applications*], there are both the 12 principles of green chemistry and of green engineering.[24] The book also states that the concept of green chemistry was "enunciated" by Anastas in 1991.

The conflict between the two visions of green chemistry was discussed by many French social scientists. Sociologist Laura Maxim explored the problem in the 2011 book *Chimie Durable: Au-delà des promesses* [*Sustainabe chemistry: beyond promises*], where she explained that terms such as green and sustainable chemistry were not understood the same way by all stakeholders.[25] However, Maxim, while interested in the critical evaluation of green chemistry promises, remained neutral on the use of the terms itself. Some scholars became nevertheless more involved and took sides. For example, the French philosopher Jean-Pierre Llored adopts in his works a clearly enthusiastic tone suggesting he sees green chemistry, as presented in the 12 principles, as the new revolutionary paradigm for

---

20 Stéphane Sarrade, *La chimie d'une planète durable* (Paris: Le Pommier, 2011).
21 Stéphane Sarrade, *Ressources de la Chimie Verte* (Paris: EDP, 2008).
22 Christine Ducamp, "Chimie verte: approche nouvelle et responsable face aux problèmes issus des activités chimiques," in *Développement durable et autres questions d'actualité – Questions socialement vives dans l'enseignement et la formation*, ed. Alain Legardez (Dijon: Educagri, 2011), 145–162.
23 Hilaire Bewa, Virginie Le Ravalec, "Le contexte: reconversion de la chimie vers la chimie verte," *OCL – Oilseeds and Fats, Crops and Lipids*,19 (2012): 1–5.
24 Jacques Augé and Marie-Christine Scherrmann, *Chimie verte: Concepts et applications* (Paris: CNRS Editions, 2017).
25 Laura Maxim, ed., *Chimie Durable: Au-delà des promesses* (Paris: CNRS Editions, 2011).

practising chemistry in general.[26] On the other hand, Martino Nieddu and his collaborators remained much more sceptical about green chemistry in its American flavour. Their publications have, in fact, very telling titles: "Existe-t-il réellement un nouveau paradigme de la chimie verte" [Is there really a new paradigm of green chemistry][27] and "La chimie verte, une fausse rupture?" [Green Chemistry, a false break?].[28] Martino Niedu's team also organised a seminar under the title "Vers une chimie doublement verte" [Towards doubly green chemistry] which reunited more than a dozen of researchers in chemistry, engineering, and economics, focused specifically on the problem of biomass and biorefineries. For Nieddu and his collaborators, green chemistry should be both environmentally-minded and bio-sourced, and they explicitly endorsed the French tradition.

How do these theoretical debates translate into the actual work of the French chemists? A quick look into the articles indexed in Scopus written by researchers affiliated with French institutions (using the same keywords as in chapter 4) does not reveal anything interesting. Catalysis, synthesis, solvents, water, biomass, and plants are the most popular keywords between 1999 and 2020, which suggests a higher than average focus on bio-sourced materials, but nothing out of ordinary. This is, however, a faulty analysis. Firstly, because Scopus does barely include any French-speaking journals and returns less than 100 papers tagged as "chimie verte." The practically-oriented studies aiming at French engineers in biorefineries are not necessarily published in English, so a substantial amount of literature may be missing from this analysis. Secondly, French chemists familiar with the problem of competing terminologies may tend to tag their work as "green chemistry" (in English) only if it fits the Anglo-Saxon definition.

A few hints on what is being actually practised as green chemistry in France were provided in a recent (2019) overview published by the French INCREASE network (International Consortium on Eco-conception and Renewable Resources).[29] The overview is entitled "Exploring the borders of a transregional

---

26 Jean-Pierre Lorred, "Towards a Practical Form of Epistemology: the Case of Green Chemistry," *Studia Philosophica Estonica* 5 (2012): 36–60; Jean-Pierre Lorred and Stéphane Sarrade, "Connecting the philosophy of chemistry, green chemistry, and moral philosophy," *Foundations of Chemistry* 18 (2016): 125–152.
27 Nieddu, Vivien, Garnier and Bliard, "Existe-t-il réellement un nouveau paradigme de la chimie verte?."
28 Martino Nieddu, Franck-Dominique Vivien, "La chimie verte: une fausse rupture? Les trajectoires d'une transition écologique," *La Découverte* 2 (2015): 139–153.
29 Marion Maisonobe, Bastien Bernela, "Exploring the borders of a transregional knowledge network. The case of a French research federation in green chemistry," 2019,

knowledge network. The case of a French research federation in green chemistry." The document refers to green chemistry because the network itself explicitly embraces the label in its official documents and, since 2013, its members have been organizing every two years the International Symposium on Green Chemistry (ISGC). The event evidently tried to connect with the global green chemistry community and its first edition in 2013 featured plenary lectures from Anastas and Clark. And yet, from the very outset of the initiative, the prime topic treated in all symposia was biomass conversion and related problems and this has been the case ever since. This line of studies is confirmed by the 2019 paper that explains that the network: "is dedicated to eco-design and renewable resources, with the aim of developing green chemistry by using biomass – a renewable carbon source – as a raw material."[30] Despite the popularization of the American definition of green chemistry in the French academia, the key institutions working on practical applications of green chemistry remain very biomass-focused. Interestingly, the advertisement for the 2022 edition of the conference calls the symposium "The world event in sustainable chemistry research & innovation;" the green part having been dropped. It appears that sustainability becomes the word of choice for the network.[31]

How to interpret the evolution of the green chemistry terminology in France? On the one hand, there is a trend that may suggest that the French chemists increasingly align themselves with the international definitions of English and American pedigree. While Colonna could publish a work entitled green chemistry without mentioning the EPA tradition in 2006, this would have been inconceivable today. Such a process would simply lead to the disappearance of the French national biomass-focused tradition with the arrival of new generations of chemists educated into the EPA paradigm. On the other hand, the INCREASE network and its symposium feature more and more prominently the concept of sustainability and do not give up on biomass at all, which may suggest a more complex underlying dynamic. In order to better understand the place of biomass, renewability, and agricultural products in the green chemistry framework, we have to have a look at the place of these concepts in the Anglo-Saxon tradition.

---

working paper hal-02053595, accessed 22/02/2022, https://hal.archives-ouvertes.fr/hal-02053595/file/MM%26BB.pdf.
30 Maisonobe and Bernela, "Exploring the borders," 4.
31 The official website of the ISGC symposia, accessed 22/02/2022, https://www.isgc-symposium.com/.

3. Biomass and renewability in the foundational texts of green chemistry (late 1990s–early 2000s)

The first interesting phenomenon relevant to the relationship between green chemistry and biomass studies at the outset of the field was identified by Estelle Garnier in her analysis of the late 1990s publications of the American Chemical Society.[32] We have seen in chapter 2 that ACS's symposia on agriculture and food-related topics were often labelled as green chemistry between 1998 and 2001. This changed only after the Green Chemistry Institute became a part of the ACS in 2001. However, neither before nor after the Institute's takeover were biomass and renewability labelled as green chemistry in the ACS publications. In the ACS green chemistry studies before 2001, agriculture was seen as a source of pollution that has to be remediated or prevented. Green chemistry was then the chemistry of making agriculture environmentally friendly. After the Institute became a part of the ACS and imposed its own understanding of green chemistry in line with the EPA's tradition, these topics disappeared from the green chemistry publications altogether. The re-contextualization of agriculture as a potential source of biomass came much later in the green chemistry literature. Of course, it does not mean that biomass and renewable feedstocks were not studied and encouraged by the American Chemical Society. On the contrary, they were simply not considered to be green chemistry at all in the 1990s and early 2000s, meaning that their authors were not familiar with the concept or did not feel they were part of the movement.

Nevertheless, Anastas and Warner's 1998 *Green Chemistry* did not completely ignore the importance of biomass. Garnier calculated that around 12% of the book was devoted to the problem of renewability and to the choice of feedstock. Not much, but the topic was certainly present. Interestingly, Garnier noted, Anastas and Warner's argumentative strategy in favour of renewable feedstock reposed on two distinct foundations. One made a connection to the notion of peak oil and to sustainability: "If our generation were to consume petroleum resources to the extent that they were no longer a viable and usable option for future generations, this would violate the goals of sustainability."[33] The other foundation concerned, however, the problem of the pollution resulting from

---

32 Garnier, "Une approche socio-économique de l'orientation des projets de recherche en chimie doublement verte," 102.
33 Paul Anastas and John C. Warner, *Green Chemistry: Theory and Practice* (Oxford: Oxford University Press, 1998), 46.

the treatment of petroleum. Anastas and Warner encouraged chemists to avoid using non-renewable feedstock because it could be more polluting than the renewable one; a perspective that was fully in line with the pollution prevention logic of the earlier EPA's works. As such, sustainability for its own sake played the second fiddle in the green chemistry's project as presented in the 1998 book.

All in all, even if absent in earlier works stemming from the EPA's tradition, biomass conversion and renewability made part of what Anastas and Warner saw as green chemistry in their 1998 work. The authors clearly showed how the use of organic waste (such as kraft black liquor or potato waste) could lead to the manufacture of various useful chemical products. They associated the topic of biomass with concepts such as biotechnology, biocatalysis, and biosynthesis, all of which they considered to be constitutive for the further development of green chemistry.

The topic of biomass and renewability of feedstocks was even more in line with Clark's version of green chemistry. He insisted on this aspect from the very beginning of his journal's existence in his editorials. In fact, in the first issue from 1999, two Dutch researchers published an article under the title "From fossil to green" in which they criticized the over-reliance on fossil fuels on the global scale and established prognoses of energy consumption for the year 2040. They called in particular for the use of biomass not just for energy but also for the manufacture of chemicals. Their article ends with an optimistic prediction:

> Scenario 2040 (almost doubling of world population, higher average standard of living, reduction of environmental pollution) seems technologically possible without the utilisation of fossil resources. … The plant will be the 'plant' of the future. … Biomass conversion to organic raw materials is able to replace the existing organic chemistry which is mainly based on fossil resources while adding structurally new materials."[34]

Sustainability and biomass are the organizing concepts of the paper that advocates in favour of abandoning non-renewable feedstocks.

However, a quick glance at other articles published in *Green Chemistry* in its early years reveals that the topic of biomass garnered almost no attention; the Dutch paper being rather exceptional. The situation looks a little bit different if instead of searching for original scientific articles published in the journal, we look at its 'news' section in which Clark and his closest collaborators presented in various information from the world of science and industry. There, the problems

---

34 Cédric Okkerse and Herman van Bekkum, "From Fossil to Green," *Green Chemistry* 2 (1999): 107–114.

of biomass and renewability are mentioned slightly more often. For example, a news section from December 1999 mentions the promise of hemp as a "green fibre source,"[35] and elsewhere, there is an article about Soya bean glue.[36]

It appears that for Clark this topic was not merely an accessory principle, but an important part of green chemistry's core identity even though his journals failed to attract specialists in the field. This failure is by no means not surprising. The biomass valorization studies were not new in any sense back in the period. While the EPA scholars could claim that the green chemistry project was novel because it focused on waste prevention and not on waste remediation, green chemistry brought nothing new to the table in terms of biomass treatment. Chemists working on this last topic had their venues, journals, and conferences, and were, if anything, more familiar with the notion of sustainability, broadly construed. And yet, Clark's ambition continued to be structuring for the field. We notably remember that in his first huge textbook from 2002 he contrasted green and brown chemistries (based on the previous work of Woodhouse).[37] Brown chemistry was brown because it was polluting, but also because it was non-renewable and fossil-based. The greenness of the green chemistry came from both its environmentally friendly character and its reliance on sustainable feedstocks. While the textbook does not have a chapter devoted exclusively to biomass, the topic resurfaces many times in its different chapters engaging with both practical and conceptual problems.

Estelle Garnier contrasted in her Ph.D. dissertation the approach towards renewability in the 1998 (Anastas and Warner) and 2002 (Clark and Macquarrie) books. She writes that for Anastas and Warner "C2V" [*chimie doublement verte* – doubly green chemistry] appears to be a compromise between sustainability and other elements of their project, while in Clark's book "it is presented as a transversal preoccupation combining with a variety of principles."[38] In other words, for Clark and Macquarrie renewability was a structuring property of all good green chemistry, not just one of its many features. Furthermore, Garnier notes one more key difference. For Anastas and Warner, biomass was supposed to be a source of building blocks replacing petroleum-based ones. In the 2002

---

35 No author, "Potential for hemp as green fibre source," *Green Chemistry*, 1 (1999): G159.
36 No author, "Soya bean glue," *Green Chemistry* 1 (1999): G3–G5.
37 James Clark and Duncan Macquarrie, *Handbook of Green Chemistry and Technology* (Oxford: Blackwell Science, 2002), 2.
38 Garnier, "Une approche socio-économique de l'orientation des projets de recherche en chimie doublement verte," 116.

book, on the other hand, there appears a more sophisticated idea concerning the place of biomass in the chemical industry. In the chapter "Green Chemistry in Practice" written by Joseph Bozell,[39] the author points out a difference between the processes that imply deep structural changes in the biomass (for example the manufacture of ethanol and lactic acid), and the ones that imply less invasive extractions (e.g. grain oil and starch polymers). This differentiation is rather implicit in Bozell's argument, but for Garnier it is a fundamental difference in chemistry philosophies, and she explores it further in the dissertation. For her, instead of trying to emulate the petroleum industry and breaking complex molecular chains into tiny bits only to reconstruct them again into final products, the truly green chemical industry should concentrate on capitalizing on the molecular complexity already present in natural materials. This approach was, in fact, also endorsed by Garnier's Ph.D. director, Martino Nieddu, who was inspired, among others, by the concept of "chimie douce" [soft chemistry] developed in the French chemical literature in the 1970s.[40]

4. The growth of the place of biomass and renewability in the literature on green chemistry

Throughout the 2000s, Clark insisted more and more on the importance of biomass and renewability. In 2006, he published one of his rare (or at least rarer than in the case of Anastas) programmatic articles outlining his vision for the development of the discipline: "Green chemistry: today (and tomorrow)." The article does not mention the 12 principles or any similar organizing concept. It insists, however, a lot on a very specific vision of the field:

> The move from our well-established petrochemical based organics chemical industry to one based on renewable resources is beginning to open the door to numerous opportunities for exciting new chemistry research including benign extraction of valuable chemicals from biomass (e.g. using supercritical fluids), adding value to nature's most abundant polymers (starch, cellulose, chitin, etc) and the bulk conversion of biomass to new "bio-platform molecules" … Bioplatform molecules are the beginning of a new and

---

39 Garnier, mistakenly, refers here to another author, Keith Martin, who wrote a chapter on biocatalysis. This does not affect her argument though.
40 See for example Nieddu's review from 2008: Martino Nieddu, "Bernadette Bensaude-Vincent et Jonathan Simon, 2008, Chemistry, The Impure Science, London, Imperial College Press, p. 268 + index.," *Développement durable et territoires* [On line], Lectures (2002–2010), 2008, accessed 22/02/2022, http://journals.openedition.org/developpementdurable/8216.

vital challenge for organic chemists. Can we build on them as we have done over the last 70 years with the now well-established petroplatform molecules such as ethene and benzene …? With the rapid growth in biotechnology we can expect to see both more candidates for new platform molecules and more selective bioprocesses for making platform molecules.[41]

Clark presents in his article numerous visual models integrating concepts such as bioplatform molecules, biorefineries, and biotechnology into a single green chemistry framework. In Clark's visions for the future of the discipline, the nature of the feedstock plays the most prominent role. Interestingly, the article is fully in line with Catalla's book published the same year. While some French chemists who published after Catalla were somewhat reluctant to embrace the French tradition and referred more willingly to the 12 principles as the foundation for what they considered green chemistry, Catalla's supposedly outmoded approach followed, in fact, the most recent trends in the mid-2000s.

The importance of biomass in green chemistry can be confirmed by the theoretical works of other scholars from the same period as well. In one of the previous chapters, I mentioned a Swedish policy analyst Jesper Sjöström who wrote in 2006 in the pages of *Green Chemistry* an article on the green chemistry policy.[42] Sjöström makes in his paper an interesting remark. At first, he establishes that green chemistry is a meta-discipline and talks about the 12 principles of green chemistry, the 12 principles of green engineering, and the 12 additional principles of Winterton. However, later he makes the following summary: "the principles of green chemistry entail: (1) renewables as chemical feedstocks, (2) substitution of hazardous chemicals, and (3) reduced consumption of chemicals and energy." Now, a careful reading of all the principles enumerated by the author certainly does not warrant mentioning the renewables as the first out of three characteristics of what green chemistry should be, unless principle 7 of GC and principle 12 of GE are blown out of proportion.[43] Of course, Sjöström was

---

41 James Clark, "Green chemistry: today (and tomorrow)," *Green Chemistry* 8 (2006): 17–21.
42 Jesper Sjöström, "Green chemistry in perspective—models for GC activities and GC policy and knowledge areas," *Green Chemistry* 8 (2006): 130–137.
43 Arguably, the principles of green engineering are more in line with renewability. For example, principle 6: Conserve Complexity ("Embedded entropy and complexity must be viewed as an investment when making design choices on recycle, reuse, or beneficial disposition") entails much of the philosophy for which argued Garnier and Nieddu. This is another proof that the 12 principles of green engineering are more thought-through and more embracing than the 12 principles of green chemistry.

fully aware of terminological difficulties and it was perfectly justifiable for him to adopt some operational definitions of green chemistry for the purpose of his work. The point is that his emphasis on renewability as the core element of green chemistry was a rather new trend that was gaining popularity over the period in theoretical debates over green chemistry's identity.

Reading Clark's or Sjöström's publications from 2006 may suggest that bioresources and biomass became one of the topics of choice for scholars publishing in the review such as *Green Chemistry*. Nothing is further from the truth. Upon looking at the number of publications featuring terms such as "biomass" and "renewable feedstock(s)," it becomes clear they were virtually absent in the green chemistry literature at that time. In fact, between 1999 and 2006 there were on average less than 10 papers yearly published with any of these terms in their titles, abstracts or keywords (Figure 21).

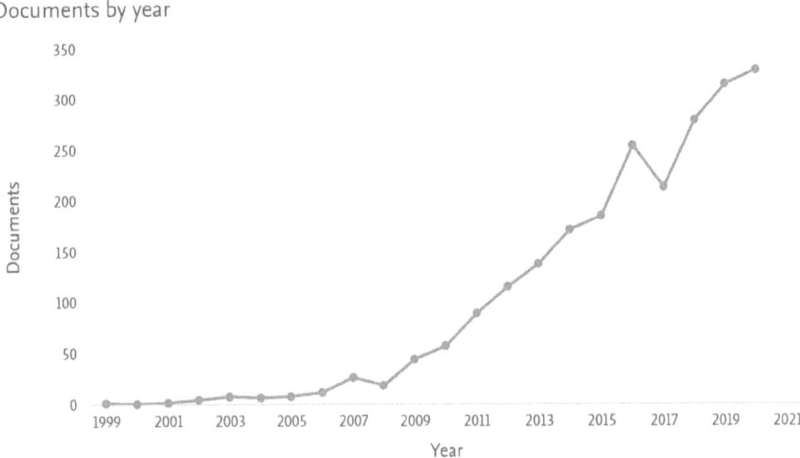

Figure 21: Number of papers with "biomass," "renewable feedstock" and "renewable feedstocks" in the titles, abstracts and keywords of green chemistry papers according to Scopus[44]

It was around 2007 that something started to change and from then on, the number of articles grew exponentially every few years, faster than the average growth of green chemistry in general. But even so, as discussed in chapter 4,

---

44 For the choice of green chemistry papers, see the methodology part of chapter 4.

neither biomass nor renewable feedstock(s) ever made it to the top five keywords and were in fact usually below the top ten. This does not represent, of course, the popularity of biomass in general in the articles indexed in Scopus. Biomass itself is the topic of thousands of articles every year, with only a fraction highlighting the connection to green chemistry (Figure 22).

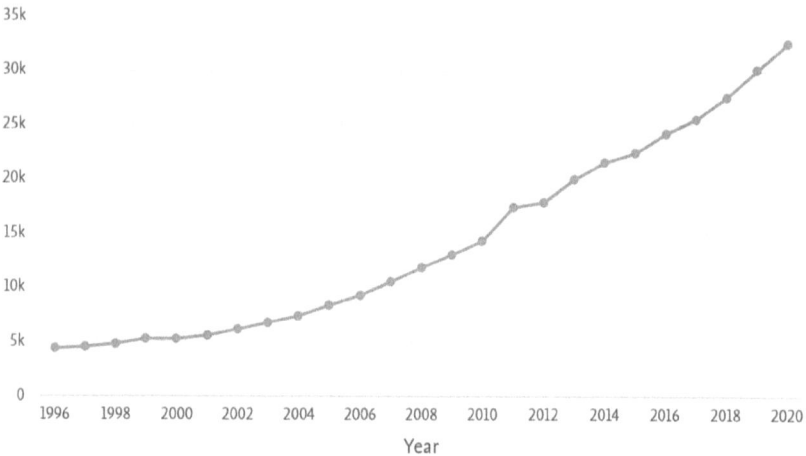

Figure 22: Number of articles published yearly with "biomass" in their titles, abstracts, or keywords according to Scopus (1996–2020)

According to these results, there would be a mismatch between the ambitions of the discipline that emphasised the importance of biomass and renewability in its programmatic theoretical works, and the actual practice of researchers. People working on biomass did not know or did not need the green chemistry label to make their work relevant. And people who identified their work as green chemistry followed a much narrower synthetic chemistry paradigm focused on pollution prevention.

However, while this might have been the reality in the early 2000s, the hypothesis becomes nevertheless much nuanced later, especially if we search the *Green Chemistry* journal for biomass-related terms (such as lignin, cellulose, agriculture, etc.), in plain text instead of in titles, abstracts, and keywords. This search returns many more results following largely the pattern from figure 21 but more prominently in terms of absolute numbers. This can lead us to another hypothesis. While some chemists labelling their work as green chemistry continued to ignore biomass altogether, for others, biomass-valorization became green chemistry *par excellence* and they did not bother highlighting it at all. An article can talk about the valorization of lignin or about agricultural wastes without using terms such as biomass or renewability. Estelle Garnier analysed 'online first' papers to be published in the *Green Chemistry* journal in 2012 and found out that around 90% of them concerned the 7th principle.[45] This appears to be an exaggeration and my own research does not confirm this estimate, but the overall tendency appears to be true. Again, the difficulty is that there is no single keyword that would allow us to evaluate these questions in a straightforward manner, and every article needs to be evaluated individually. Not to mention the fact that biomass, biocatalysis, or similar topics are often featured alongside other green chemistry themes; they may be of secondary importance in an article focused on a different topic (e.g. manufacturing of ionic liquids). Nevertheless, my estimate is that between 2010 and 2020, the biomass-related problems occupied between 25%-50% of the volume of *Green Chemistry* depending on the choice of keywords (for the full-text research).

Whatever the exact proportions are, what matter is that while biomass and renewability were barely visible topics in green chemistry publications in the early 2000s, they rose to prominence in the second half of the 2000s and it has not changed since. Clark and his colleagues rejoiced over this development in a 2014 editorial celebrating the 15th anniversary of the journal:

> A cursory examination of recent issues of Green Chemistry reveals another important development of the last 15 years: the advent of the bio-based economy. The utilisation of biomass as a renewable feedstock for the production of biofuels, commodity chemicals and greener, biomass-based materials, such as bioplastics, affords substantially reduced environmental footprints compared with analogous products derived from finite fossil resources.[46]

Biobased economy, bioeconomy, or biotechonomy is another eclectic term that became increasingly popular throughout the 2010s to capture the complexity of the issues surrounding sustainability and renewability-related problems. The green chemists jumped on its bandwagon just in time (Figure 23).

---

45 Garnier, "Une approche socio-économique de l'orientation des projets de recherche en chimie doublement verte," 130.
46 James Clark, Roger Sheldon, Colin Raston, Martyn Poliakoff and Walter Leitner, "15 years of Green Chemistry," *Green Chemistry* 16 (2014): 18.

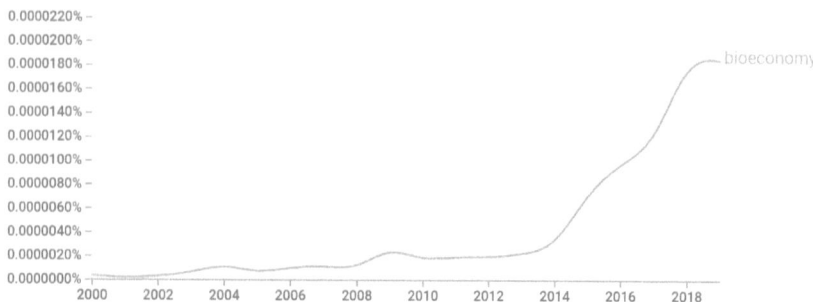

Figure 23: The popularity of the term bioeconomy according to Google Ngram (2000–2019)

Also in 2014, Roger Sheldon, the creator of the E-factor, published in *Green Chemistry* an overview on the topic of biomass: "Green and Sustainable Manufacture of Chemicals from Biomass: State of the Art" in which he explained:

> Although attention [in green chemistry publications] continues to be focused on waste minimisation and avoiding the use of toxic and/or hazardous reagents and solvents … the last decade has witnessed a growing emphasis on the third element of green chemistry, namely the substitution of non-renewable fossil resources – crude oil, coal and natural gas – by renewable biomass as a sustainable feedstock for the manufacture of commodity chemicals and liquid fuels. Indeed, the ultimate in sustainable development must surely be to emulate nature by harnessing the sun's energy for the synthesis of chemicals from carbon dioxide and water. A so-called biobased economy is envisaged in which biofuels, commodity chemicals and novel materials such as bioplastics will be produced in integrated biorefineries.[47]

The passage above appears to be the new credo of green chemistry in the second half of the 2010s.

This shift towards biomass became even more pronounced in later more theoretical and systematic papers on what green chemistry should be. In 2019, *Green Chemistry* published a special issue on "Green Biorefinery Technologies based on Waste Biomass" featuring 21 original articles on the topic.[48] Clark's editorial for the issue expresses the new philosophy of the field:

---

47 Roger Sheldon, "Green and Sustainable Manufacture of Chemicals from Biomass: State of the Art," *Green Chemistry* 16 (2014): 950–963.
48 James Clark, "Green biorefinery technologies based on waste biomass," *Green Chemistry* 21 (2019): 1168–1170.

While there seems to be a real need for people to review their attitudes about resources and the value of articles, a successful transition from a linear to a more circular economic model must be driven by industry. This must be based on the use of renewable resources, green manufacturing and the production of environmentally compatible, recyclable products. Since the most widely useful element in society is carbon and the most widely used resource is fossil carbon, it seems appropriate that we place a lot of attention on more sustainable organic products and on their production from biomass. The choice of biomass is critical – being renewable does not guarantee being green and we must learn from our experience of biofuels that the sourcing of the biomass is critical. The use of non-food biomass is vital and even better if we focus on bio (mass)-waste. These can include agricultural waste, food processing waste, municipal solid waste and sewage sludge, although the organic components in these waste streams may not be entirely bio-based, with (petroleum derived) plastic being an increasingly common component of waste. By using bio-waste feedstocks to make bio based products, we can help solve both resource and waste problems and be "double green."

Bio-waste, circular economy, environmentally compatible products... The buzzwords describing different aspects of sustainability continue to multiply in the recent literature on green chemistry. Circular economy is perhaps the most prominent one as it even spurred the concept of circular chemistry I discuss later. Most notably, Clark uses the expression "double green chemistry." While it remains unclear whether the term "double green" was directly or indirectly inspired by the French scholarship on green chemistry, it covers broadly the same idea as in France. Green chemistry should be environmentally friendly, but to do so it must rely on renewable feedstocks. Interestingly, Clark's editorials constitute a solid proof that the French green chemistry tradition was not some peripheral outlier stemming from a backward chemical culture lagging behind the American or British one. On the contrary, it appears that Catalla's 2006 book better anticipated big trends in the green chemistry philosophy of the early 2020s than some later French manuals that tried to link green chemistry to the early EPA's pollution prevention line of thinking.

Another result of this shift towards biomass also affects the reconstruction of green chemistry's genealogy. After all, if the environmentally-friendly biomass treatment is at the heart of the green chemistry community today, the pioneering Italian *Chimica verde* from 1987 should be considered the founding book of the discipline and Pasquon and Zanderighi its true fathers.

At the same time, one could reasonably argue that green chemistry was the EPA's pollution prevention philosophy, and what we witness today is something different altogether. It may be asked whether we did not reach the limits of what can be described as green chemistry and whether some sort of new terminology, for example "sustainable chemistry," should not come to substitute it.

Interestingly, many such competing theoretical frameworks have been considered over the last decades.

Chapter 7: Not only green: sustainable chemistry and past environmentally-friendly chemistries

So far I have followed the history of the term "green chemistry" in its various iterations and explained how it came to dominate the language of environmentally-minded chemists, especially with simple codes of conduct such as the 12 principles. However, what I have not yet studied are alternative conceptualisations, terminologies, and codes of conduct, stemming from different intellectual traditions and using different names. Some of them became largely forgotten, some did not manage to gain the same traction as green chemistry, others are gaining steam only right now.

This chapter is divided into two sections. In the first one, I study the past alternatives: the European, and in particular German, forerunners of green chemistry. The formalization of the idea of pollution prevention in Germany antedates much of the EPA's work in the 1990s. This German influence carries on to the second section of this chapter about the emerging concept of sustainable chemistry which is, if not independent, at least autonomous in its relationship to green chemistry.

A word of caution is necessary: in this chapter, I summarize some key ideas and tendencies, but I do not provide their exhaustive analysis. For example, my short introduction to the concept of industrial ecology does not give justice to this rich discipline with its own journals and textbooks, but its history must be briefly presented to explain the relationship between the green chemistry community and sustainability scholars. Similarly, the sections on the concepts such as soft chemistry are not here to provide their thorough analysis, but merely to affirm their place in the history of environmentally friendly tendencies chemistry, precisely because they are notoriously absent in the historical publications written so far. In other words, this chapter is tasked with mapping ideas and with providing the compass to those interested in digging deeper.

1. Forgotten alternatives

Before the rise of green chemistry, many attempted to formalize new ways of thinking about the relationship between chemistry and the environment as well as about pollution prevention. Some of these chemistries, while originally ambitious, devolved into something much narrower. Others lacked proper institutionalization or were politically problematic. The message of the section is that it takes more to win over the hearts of scientists and policymakers than just to have a good idea.

It is important to note that there were a plethora of such competing conceptualisations. Of course, not every single idea deserves the same attention. Some of them were too narrow or not enough formalised, whereas others were too broad. For example, an overview of green and sustainable chemistry written in 2001 by a group of Dutch scholars mentioned two concepts as if they were on the same level as green chemistry: cascade chemistry and design for the environment.[1] As for cascade chemistry, the authors explain that it "promotes the synthesis of compounds without very isolating intermediate products and by using the mildest possible conditions: ideally a temperature of 258°C, atmospheric pressure, and an environmentally safe solvent such as water, methanol, or acetone." The concept itself was first proposed in a 1994 *Science* article on one-pot syntheses, which, interestingly, also mentioned the term green chemistry to describe the type of research presented in its pages.[2] One-pot syntheses became widely labelled as green chemistry and there was little in "cascade chemistry" itself that would make it sufficiently different to form a separate tradition. The term was used sparsely in the following decades and was largely absorbed by green chemistry. As for Design for the Environment (DfE), it is a large inclusive concept that, while not as popular today as it used to be in the 1990s when the EPA officially endorsed and promoted it, still functions as a benchmark for sustainability scholars. However, it does not concern exclusively chemistry, it connects to a wider literature on sustainability, and remains only loosely formalized. It is rather an overarching guideline than a proto-discipline or a paradigm that green chemistry aspires to be. The point is that while there were many ideas that circulated in the 1990s that created a climate prolific for the development of green chemistry, it is hard to compare them or discuss them within the same framework. In this section, I exclude these various concepts and concentrate instead

---

1 André van Roon, Harrie A.J. Govers, R. John and Hans van Weenen, "Sustainable chemistry: an analysis of the concept and its integration in education," *Journal of Sustainability in Higher Education*, 2 (2001): 161–180.

2 Nina Hall, "Chemists clean up synthesis with one-pot reactions," *Science*, 266 (1994): 32–34.

on a selection of genuine alternatives that share, at least to some extent, similarities with green chemistry.[3]

1.1. Solid state-chemistry with green ambitions: French chimie douce

Soft chemistry or "chimie douce" (the term appears in French in many English publications) is a concept initially developed in the late 1970s. It is sometimes defined as the solid-state chemistry in ambient temperatures.[4] Two chemists, Jacques Livage and Jean Rouxel, are usually cited as its founding fathers, representing two distinct academic traditions in the French solid-state chemistry. While a typical solid-state chemistry requires high pressures and high temperatures, Livage argues that:

> Living organisms have developed strategies completely different from those used by our materials engineers. Silicate glasses, for example, are made by melting silica sand above 1000 °C, whereas diatoms and radiolarians build sophisticated silica structures at room temperature directly from the very small amount of silica dissolved in seawater.[5]

In other words, according to Livage, soft chemistry aims at the reduction of energy necessary for chemical reactions, but it also comes with an overarching ambition to emulate natural processes. Some articles that refer to the concept present chimie douce as the synthesis of biomimetic and bioinspired materials in a manner imitating the 'manufacturing strategies' deployed by living organisms.[6] Chimie douce also was, as I have already mentioned, an important inspiration for Martino Nieddu's and Estelle Garnier's further work on doubly green chemistry.[7]

---

[3] Some topics discussed in this section have been presented in a slightly different light in: Marcin Krasnodębski, "Lost green chemistries: history of forgotten environmental trajectories," *Centaurus* (accepted).

[4] Pierre Teissier, "Une histoire de la chimie du solide" (PhD diss. University of Nantes, 2008), 369.

[5] Jacques Livage, "Chimie douce: from shake-and-bake processing to wet chemistry," *New Journal of Chemistry* 25 (2001): 1.

[6] Markus Antonietti and Maria-Magdalena Titirici, "Coal from carbohydrates: The 'chimie douce' of carbon," *Comptes Rendus Chimie* 13 (2010): 167–173; Clement Sanchez, Laurence Rozes, François Ribot et al., "'Chimie douce:' A land of opportunities for the designed construction of functional inorganic and hybrid organic-inorganic nanomaterials," *Comptes Rendus Chimie* 13 (2010): 3–39.

[7] Martino Nieddu, "Bernadette Bensaude-Vincent et Jonathan Simon, 2008, Chemistry, The Impure Science, London, Imperial College Press, p. 268 + index.," *Développement*

The foundational article of the field was an editorial piece by Jacques Livage in the French nation-wide newspaper *Le Monde* from 1977.[8] Its title was "Towards an ecological chemistry. When air and water replace petroleum" and it explored common environmental themes such as the growing distrust towards chemistry, the inevitable paucity of raw materials (such as fossil fuels), and the problem of wastes and chemical pollution. The term chimie douce was used there only once in the phrase: "Are we, however, doomed to observe the growth of wastefulness and of pollution, or will it be possible to invent a "soft" chemistry more harmoniously integrated to the natural processes?"

And yet, even if the overarching ambition and the catchy term itself were effectively laid down in the 1970s, the concept itself developed along much narrower lines. The term chimie douce has been above all used to describe a strand of solid-state chemistry, concentrating on very specific topics such as, for example, sol-gel processes. The majority of chemists referring to the concept mean it in a circumscribed way without connecting it to the broad environmental ambitions of Jacques Livage at all. Livage admits himself that the foundational article is not representative of chimie douce as it developed afterwards, but he continues to see it as the basis for the concept anyway.[9] The fact, that the chimie douce had another founding father, Jean Rouxel, in the works of whom the environmental dimension was much less explicit, makes this evolution understandable.

As a consequence, from the perspective of its environmental ambition, the overall impact of chimie douce appears somewhat ambiguous. The concept itself is alive and well but remains understood in a narrow manner. This is not to deny that chimie douce can be in line with the green philosophy, but one can ask whether it could not have gone in a more general direction beyond some applications of solid-state chemistry and become for the French what green chemistry became in the English-speaking world. It may be objected that the 12 principles of green chemistry are much more all-embracing, but it should not be forgotten that Livage's environmental concerns were very broad as well. Indeed, Livage continued to insist on biomimetics throughout his entire career well into the late

---

durable et territoires [On line], Lectures (2002–2010), 2008, accessed 22/02/2022, http://journals.openedition.org/developpementdurable/8216.

8   Jacques Livage, "Vers une chimie écologique. Quand l'air et l'eau remplacent le pétrole," *Le Monde*, 26 October, 1977.
9   See for example: Jacques Livage's inaugural speech from 2001 published in Richard-Emmanuel Eastes and Bernadette Bensaude-Vincent, ed., *Philosophie de la chimie* (Paris: De Boeck Superieure), 135.

2000s, giving the concept an original and interesting spin somewhat lacking in green chemistry.[10] If green chemistry could expand from specialised studies on more efficient catalysts and synthetic pathways into a much broader overarching chemistry of sustainability, the work of the French chemists could have taken a similar direction. As such, chimie douce is an interesting episode in the history of chemistry in France, but nothing beyond that (from the sustainability point of view at least).

1.2. Politically incorrect green chemistry: German sanfte Chemie

The French soft/gentle chemistry (chimie douce) should not be confused with the German soft/gentle chemistry (sanfte Chemie), as they do not share the same genealogy. If the French concept was successful but remained limited to a relatively narrow speciality, its German equivalent became largely forgotten but was considerably more ambitious. Some of its key architects were the biologist Arnim von Gleich and the chemist Hermann Fischer, who developed their ideas in the 1980s and 1990s.[11]

The German sanfte Chemie shares some similarities with the French one. One paper defines it in the following way:

> In the German concept of soft or gentle chemistry ("sanfte Chemie"), nature is considered to be leading far ahead with respect to the development of substances and products, from which much can be learned in the process of making economic activities compatible with natural cycles ... It promotes the use of the sun, of renewable resources and biological processes. Gentle chemistry is strongly inspired by nature, which makes it distinct from green chemistry.[12]

The inspiration by nature was, at least in theory, an inspiration for chimie douce as well. However, the Germans pushed the concept much further.

Of course, some German chemists in the late 1980s understood the term "sanfte Chemie" in a relatively narrow way, as the chemistry conducted at normal atmospheric pressure, room temperature and with Ph 7. This is not what the

---

10   Jacques Livage and Thibaud Coradin, "Le verre biologique inspire les chimistes," *Pour la science* 371 (2008): 30–34.
11   Hermann Fischer, *Plädoyer für eine Sanfte Chemie* (Braunsweig: Allembik Verlag, 1993); M. Kirschner, "Zauberstoff für eine Sanfte Chemie," *Bild der Wissenschaft* 4, 1993: 14–18.
12   André van Roon, "Designing sustainable chemicals: predictive tools for the environmental fate of monoterpene pesticide" (Ph.D. diss. University of Amsterdam, 2006), 52.

partisans of the concept meant though. "The inventors of the term "gentle chemistry" understand by it 1. a different material basis, i.e. switching to natural raw materials, as little transformed as possible, and 2. a different, gentler form of knowledge, i.e. a gentler and holistic chemical Science." explained an article from 1988.[13] The idea that chemistry should not merely switch to more natural and less-transformed materials, but that our entire thinking about the relationship with nature should drastically change is perhaps one of the most interesting contributions of sanfte Chemie's supporters.

Unlike green chemistry that developed within the walls of the federally funded EPA and hoped to peacefully convince the chemical industry that it should adopt its ways, sanfte Chemie was explicitly political from its very inception. Both Gleich and Fischer, along with many other pioneers in the field, were activists and supporters of the German green party (Bündnis 90/Die Grünen) and worked on the concept with its benediction in a special working group. Sanfte Chemie was to be the core concept behind the party's Chemiepolitik and to replace the old policy that targeted, above all, chlorinated hydrocarbons. The working group's members contributed to studies on sanfte Chemie from the technological, scientific, and even philosophical point of view.

Two essential publications in the field are Gleich's *Der wissenschaftliche Umgang mit der Natur. Über die Vielfalt harter und sanfter Naturwissenschaften* [*The scientific approach to nature. About the variety of hard and soft natural sciences*],[14] which discussed the broader epistemological challenges concerning the place of science in society, and Fischer's *Plädoyer für eine Sanfte Chemie* [*A plea for soft chemistry*] focused specifically on soft chemistry itself.[15] The ambition of both authors was to completely and thoroughly transform the way science is practised in favour of less invasive and holistic ways of knowing. Unlike green chemistry, which insisted on its own novelty in the 1990s and was loosely tied to the American environmentalist tradition, sanfte Chemie was grounded in abundant and eclectic literature in the history and philosophy of science, environmentalism, but also German Marxism (e.g. Herbert Marcuse and Ernst Bloch) and even mysticism (e.g. Rudolf Steiner). Fischer went as far as to link his approach to Goethe's philosophy of science.

---

13 H. Müller, "Sanfte Chemie," *Nachrichten aus Chemie, Technik und Laboratorium* 36 (1988): 1011.

14 Arnim von Gleich, *Der wissenschaftiche Umgang mit Natur – Über die Vielfalt harter und sanfter Naturwissenschaften* (Frankfurt: Campus Verlag, 1989).

15 Fischer, "Plädoyer für eine Sanfte Chemie," 21–24.

In terms of formalization, it was Hermann Fischer who conceived nine core theses of sanfte Chemie which succinctly summarized the main characteristics of the concept (Table 25).

Table 25: Core Theses of Sanfte Chemie (soft/gentle chemistry). My clarifications in square brackets based on Fischer's comments.

---

1. Analysis of the entire product biographies [that can be understood as a comprehensive LCA with a particular focus on ecotoxicology]
2. Use of nature as a model for the synthesis of materials [e.g. biomimetics]
3. Use of diversity and complexity
4. The violence against the substances turns against us [Fischer calls for a new ethics for dealing with the inanimate world]
5. Avoid interference with natural structures
6. Solar energy is an optimal source of energy
7. Use of plant-based chemistry that poses no accident risk [For Fischer, imitating nature and using plant-based chemicals is inherently safer that the current practices of the chemical industry]
8. Hazardous waste can be avoided
9. Gentle chemistry closes material cycles [i.e. it advocates renewability and is in line with industrial ecology principles]

These nine theses are somewhat less elegant than the 12 principles of green chemistry from 1998, but some of them go far beyond them. They include safe syntheses, life cycle assessment, conservation of complexity, renewability of feedstocks and energy, and biomimetics. These theses accommodate well very recent trends in green chemistry such as the synthesis of nanoparticles with plants and the circularity of material use. Sanfte Chemie was, in many respects, truly a green chemistry *avant la lettre*.

While Woodhouse and Clark argued that there was a divide between the brown and green chemistries, for sanfte Chemie proponents, the divide was between the hard and soft sciences. But while the green/brown divide was focused on raw materials and toxic pollution, the hard/soft divide was about the entire way of interacting with nature: one asserting power over it, the other harmonious and conciliatory. It is also important to note that, unlike the American scholars who gave the impression that they were inventing things from scratch and revolutionizing chemistry, Fischer devoted a full chapter to chemurgy in the US in the 1930s and was aware of similar environmental lines of thoughts in the past.

Obviously, Fisher's and Gleich's books had openly political and ideological undertones and were in favour of radical green policies, which might have led to the project's overall failure in the long run. One of the reviewers of Fischer's work, critical of the new ethics on inanimate matter advocated by the German chemist, asked mockingly in 1994 whether the book really put into question "400 years of the history of science, the development of human knowledge, the triumphant advance of Western thought."[16] In fact, whether Fischer went too far in his rejection of the traditional chemistry is an open question. The underlying assumptions behind sanfte Chemie deserve certainly a thorough critical reassessment. The point is, however, that many of its elements anticipated what would become seriously considered by green chemists over the last decades.

Overall, the glory years of the concept were the late 1980s and early 1990s, but it fell later into disuse and did not reach a wider public. It appears there was only one major governmental project explicitly referring to the sanfte Chemie terminology funded in Austria in 1997 under the title "Nachwachsende Rohstoffe und Sanfte Chemie" [renewable raw materials and gentle chemistry], but the term was understood there in a somewhat narrower way simply as sustainability, renewability, and conservation of complexity.[17] Unsurprisingly, some German sources simply cite it today as another name for green and sustainable chemistry without differentiating these various concepts.[18] This is not wrong *per se*, the sanfte Chemie theses satisfy the criteria for sustainability and greenness. However, it should not be forgotten that sanfte Chemie has an independent genealogy preceding much of the conceptual work in green chemistry, and that its theses were much more radical.

Even though sanfte Chemie failed to gain more traction, that does not mean that the core ideas behind it were abandoned. Arnim von Gleich continued to work on sustainability, biotechnology, and biomimetics for many years, and Hermann Fischer remains a renowned manufacturer of eco-friendly pigments. Fisher has recently (in 2017) published a book on the transformation of chemistry, *Chemiewende*,[19] and maintains an on-line activity regularly publishing on

---

16 Karl-Geert Malle, "Sanfte Chemie halbokkult?," *Nachrichten aus Chemie, Technik und Laboratorium* 42 (1994): 64.
17 Report "Mit Nachwachsenden Rohstofen auf dem Weg zur Nachhaltigkeit," *Forschungsforum* 3 (1997), accessed 22/02/2022, https://nachhaltigwirtschaften.at/resources/nw_pdf/fofo/fofo3_97_de.pdf.
18 Website of the German environmental think tank Katalyse Institute, accessed 22/02/2022, http://umweltlexikon.katalyse.de/?p=1115.
19 Hermann Fischer and Horst G. Appelhagen, *Chemiewende: Von der intelligenten Nutzung natürlicher Rohstoffe* (Antje Kunstmann, 2017).

issues relating to the environment, sustainability, and biodiversity in the context of chemistry.[20]

In general, the failure of the sanfte Chemie conceptualisation may result from its radical nature. If we were to treat sanfte Chemie seriously, we would have to reshape the entire industrial ecosystem to emulate the natural processes. If this idea is perhaps becoming less unthinkable today, it certainly was in the early 1990s.

1.3. Alternative pollution prevention frameworks in Europe

While the holistic environmental concept of sanfte Chemie might have been too much for mainstream chemists in the 1990s, it would be unfair to paint the German chemical industry in black and white colours and present it as completely unaware of environmental challenges. The problem of pollution prevention had been widely discussed by German industrial chemists long before the American green chemistry took over. One of the first German books advocating some form of pollution prevention, or at least taking environmental considerations into account in manufacturing, was published already back in 1978.[21] By the early 1990s, the topic of pollution prevention flourished in German-speaking publications and there were at least a few attempts to formalise the benign-by-design philosophy in a way completely independent from the ones conducted in the US.[22]

To be clear, one such attempt was not exclusively German but a pan-European one, namely the 1996 European Union Council Directive 96/61/EC. So far, I refrained from discussing legislation in this book. Undoubtedly, legislation plays a primordial role in making chemical products safer. The 1990 Pollution Prevention Act was an important impulse for the development of the EPA's green

---

20 Official website of the company AURO, accessed 22/02/2022, https://www.auro.de/de/ueber-AURO/sanfte-chemie/fachbeitraege/.
21 Cristoph Lange, *Umweltschutz und Unternehmensplanung. Die betriebliche Anpassung an den Einsatz umweltpolitischer* (Wiesbaden: Gabler KG, 1978).
22 D. Becher, "Vermeiden, Vermindern, Verwerten – integrierter Umweltschutz in der Produktion," in the report *Die Bayer-Umweltperspektive II* (Leverkusen, 1991), 34; G. Scharfe, B. Seweko, "Produktionsintegrierter Umweltschutz – Das Beispiel Bayer," *Chemische Industrie, Sonderausgabe Nordrhein-Westfalen* (1991): 17–20; Claus Christ, "Umweltschutz in der chemischen Industrie – Vermindern und vermeiden von Abfallen," in *Umwelt, Logistik und Verkehr*, ed. Reinhardt Junemann (Dortmund: Praxiswissen GmbH, 1992).

chemistry philosophy. Similarly, the famous REACH (Registration, Evaluation, Authorisation and Restriction of Chemicals) regulation from 2006 had a tremendous impact on the practices of the chemical industry in Europe. But while the chemical legislation is about preventing dangerous chemicals from being manufactured or circulating in the environment, it does not provide guidelines on how to make better, greener, and more sustainable chemistry. It is up to the industry itself to develop strategies and protocols to avoid violating existing rules. In principle, the 96/61/EC directive is no different in this respect. It concerns "integrated pollution prevention and control" and "lays down measures designed to prevent or, where that is not practicable, to reduce emissions in the air, water and land from the above mentioned activities, including measures concerning waste, in order to achieve a high level of protection of the environment taken as a whole." It has the same ambitions as the Pollution Prevention Act in the US. Since the European directives are not laws strictly speaking and they have to be implemented by the national legislatures, its mission was to create a system of information exchange, similar rules, and similar vocabularies to address the issue of pollution prevention on a pan-European level.

The reason why I mention the directive is that throughout the text it regularly refers to the concept of "best available techniques" which are defined as

> the most effective and advanced stage in the development of activities and their methods of operation which indicate the practical suitability of particular techniques for providing in principle the basis for emission limit values designed to prevent and, where that is not practicable, generally to reduce emissions and the impact on the environment as a whole.

To grossly simplify, the EU member-states are invited in the directive to attribute permits for chemical plants only if they apply "best available techniques" to prevent pollution. If this sounds vague, an inconspicuous annex IV of the directive provides a list of "Considerations to be taken into account ... when determining best available techniques ... bearing in mind the likely costs and benefits of a measure and the principles of precaution and prevention" (Table 26).

Table 26: Rules for determining best available techniques in Annex IV to Council Directive 96/61/EC

> 1. The use of low-waste technology;
> 2. The use of less hazardous substances;
> 3. The furthering of recovery and recycling of substances generated and used in the process and of waste, where appropriate;
> 4. Comparable processes, facilities or methods of operation which have been tried with success on an industrial scale;
> 5. Technological advances and changes in scientific knowledge and understanding;
> 6. The nature, effects and volume of the emissions concerned;
> 7. The commissioning dates for new or existing installations;
> 8. The length of time needed to introduce the best available technique;
> 9. The consumption and nature of raw materials (including water) used in the process and energy efficiency;
> 10. The need to prevent or reduce to a minimum the overall impact of the emissions on the environment and the risks to it;
> 11. The need to prevent accidents and to minimize the consequences for the environment;
> 12. The information published by the Commission pursuant to Article 16 (2) or by international organizations

These, incidentally 12, considerations were to be at the heart of the European pollution-prevention policy. Their strong points and inadequacies were thoroughly discussed in the policy and management literature.[23] What is interesting to us are the parallels with the green chemistry principles. Some of these considerations are purely formal (4, 5, 6, 7, 12), but others are much more green-oriented strictly speaking. In fact, considerations 1, 2, 9, 10, and 11 can be found in a similar form among the 12 principles of green chemistry. Principle 3, recycling, goes even beyond the boundaries of the green chemistry in the traditional sense.

What differentiates the EU rules from the EPA's 12 principles is their target audience. The former were directed to regulators and evaluators and not to the 'ground troops' of chemistry, unlike the latter, which became the bread and butter of academic and industrial chemists. Indirectly, however, the European principles certainly introduced the conceptual grid that facilitated the spread of green

---

23 Valérie Laforest, "Applying Best Available Techniques in Environmental Management Accounting: From the Definition to an Assessment Method," in *Environmental Management Accounting for Cleaner Production*, ed. Stefan Schaltegger, Martin Bennett, Roger L. Burritt and Christine Jasch (Springer, 2008): 29–47, 35.

chemistry ideas later on. Interestingly, a recent (2021) article on the history of green and sustainable chemistry places the EU directive's rules at the same level as the 12 principles, as if the two codifications were elements of a single environmental trajectory.[24] This is true in a very broad sense, but from the historical perspective, their genealogies and audiences were very different.

What is striking is that the 1990s were a golden age of this type of codes of conduct. In 1993, the sanfte Chemie principles were formulated, in 1996, the EU pollution prevention ones, and in 1999 another list was constructed by the German chemist Claus Christ who presented his ideas in the English-speaking book *Production-Integrated Environmental Protection and Waste Management in the Chemical Industry*.[25] The concept of Production-Integrated Environmental Protection was according to Christ solidly rooted in the German scientific literature and he placed it explicitly in the environmental trajectory initiated by the German chemical industry in the 1970s.[26] He notably invoked the Association of the German Chemical Industry as the institutional backbone for this line of thinking. According to Christ, the reduction and prevention of waste could be achieved through 8 points (Table 27).

---

24 Vânia G. Zuin, Ingo Eilks, Myriam Elschami and Klaus Kümmerer, "Education in green chemistry and in sustainable chemistry: perspectives towards sustainability," *Green Chemistry* 23 (2021): 1594–1608.
25 Claus Christ, *Production-Integrated Environmental Protection and Waste Management in the Chemical Industry* (Weinheim: Wiley-VCH, 1999).
26 Christ, *Production-Integrated Environmental Protection*, 10.

Table 27: Eight Rules of Christ (for Production-Integrated Environmental Protection)

> 1. Improving the chemical process with the aid of new synthesis routes; e.g. in the production of aromatic amines, chemical reduction with iron chips is replaced by catalytic reduction by hydrogen.
> 2. Shifting the equilibrium. The use of more favourable reaction conditions can cause the position of the equilibrium to be shifted so that one of the two components A or B is almost 100% reacted. This can be achieved by using the second component in excess, by removing the product, or by using a more favourable temperature or pressure.
> 3. Improving selectivity. A very effective method of reducing the amounts of residues and improving the yield is to increase the selectivity of the chemical reaction. Examples of this include the following: Improvement of the selectivity of catalysts, e.g. by using catalysts that lower the rate of undesired side reaction, maintenance of high catalytic activity, e.g. by avoiding contact poisons, optimization of reaction conditions, e.g. by utilization differences in the reaction kinetics of the main reaction and the side reaction, more favourable temperature profiles and residence times, or more suitable reactors, recycling of the side products (if the side reaction is reversible).
> 4. Developing new catalysts, e.g., in production of polypropylene without generation of wastewater using improved metal-organic catalysts.
> 5. Process optimization.
> 6. Changing the reaction medium. If water is replaced by an organic solvent in syntheses, contamination of waste water can often be drastically reduced. However, environmentally friendly solvent handling involves not only recovery of the solvent from liquid media but also prevention of losses to the atmosphere during storage transport, production, and subsequent processing. This can be achieved e.g. by adsorptive recovery of solvents from the gas stream.
> 7. Using raw materials of higher purity.
> 8. Replacing or eliminating auxiliaries that have a harmful effect on the environment (e.g. chlorinated hydrocarbons).

Perhaps somewhat less concise than the 12 principles of green chemistry, the 8 rules of Christ for achieving Production-Integrated Environmental Protection offer nevertheless a comprehensive strategy not only for decision-makers but also for chemists and engineers seeking simple rules of thumbs for developing new products and processes. At no point in Christ's book, does he mention the EPA, Anastas, or green chemistry for that matter. It appears that the 8 rules were almost exclusively the fruit of the long development of sustainability-related industrial reflection in Germany.

In one 2020 publication on the history and state of art of green chemistry, Christ's rules were placed at the same level as the 12 principles of green chemistry and the two codifications were presented as the canon of sustainable thinking in chemistry.[27] Just as in the previously mentioned article comparing the 12 principles with the 12 EU pollution prevention recommendations, there appears to be in these and other similar publications a certain level of frustration that these alternative codifications, while roughly equivalent to the green chemistry principles, never gained a fraction of their popularity. These alternative trajectories are being rediscovered and reconstructed in order to revise the dominant narrative and justify the sense of autonomy of mainly European researchers who may not fully embrace the EPA's green chemistry. Many of these forgotten concepts have been later repurposed in particular to serve green chemistry's most dangerous rival: sustainable chemistry.

2. Escaping the green: sustainable chemistry

Sustainability is a generic keyword applicable to almost all concepts I referred to in this work. It transgresses hard sciences, economics, social sciences, culture, art, and other spheres of human life. The books on sustainability are plentiful and so are its definitions. The famous Brundtland report "Our Common Future" from 1987 explains that: "Sustainable development is development that meets the needs of the present without compromising the ability of future generations to meet their own needs," but the problem of what sustainability means in practice, and whether sustainability and sustainable development are genuinely sister concepts, remains controversial.[28] It is also important to underline that while the term itself gained popularity in the 1980s, in the German-speaking world, the equivalent word "Nachhaltigkeit" has much deeper roots and stretches back to the eighteenth century in the context of forestry.

The history of sustainability has been a subject of a few studies offering different takes on the topic.[29] The concept's genealogies usually mention Rachel Carson, environmental concerns of the 1970s, chemical and industrial accidents,

---

27 Dieter Lenoir, Karl-Werner Schramma, Joseph O. Lalah, "Green Chemistry: Some important forerunners and current issues," *Sustainable Chemistry and Pharmacy* 18 (2020): 100313.

28 World Commission on Environment and Development, "Our Common Future," accessed 22/02/2022, https://sustainabledevelopment.un.org/content/documents/5987our-common-future.pdf.

29 Ulrich Grober, *Sustainability: A Cultural History* (Totnes: Green Books, 2012); Jeremy L. Caradonna, *Sustainability: A History* (Oxford: Oxford University Press, 2014); Jeremy L. Caradonna, *Routledge Handbook of the History of Sustainability* (New York: Routledge, 2018).

and the international activism to address these issues. In different words, all these provisional narratives share many common elements with the genealogies of green chemistry, but also of toxicology, industrial ecology, environmental chemistry, and many other more or less defined disciplines and concepts. Someone without a proper historical background may perceive these various frameworks and modes of thinking as offshoots of the overarching and all-embracing idea of sustainability. This is, however, not true. It is extremely important to remember their distinct genealogies. Every single of these ideas is rooted in some tradition or in some previously existing trajectory, with its own particularities, often older than the concept of sustainability itself. That is why I reluctantly referred to sustainability in the previous chapters, despite the fact that the stakeholders themselves used the term abundantly. For example, both in the French and English publications on biomass in the context of green chemistry, the term sustainability was omnipresent. In fact, their authors regularly conflated the vocabularies of renewability, biomass, and sustainability. However, the topics such as the renewability of feedstocks and the use of biomass precede the emergence of sustainability as a separate concept; the chemical studies on these topics did not need the idea of sustainable development or sustainability to successfully grow in terms of popularity (Figure 24).

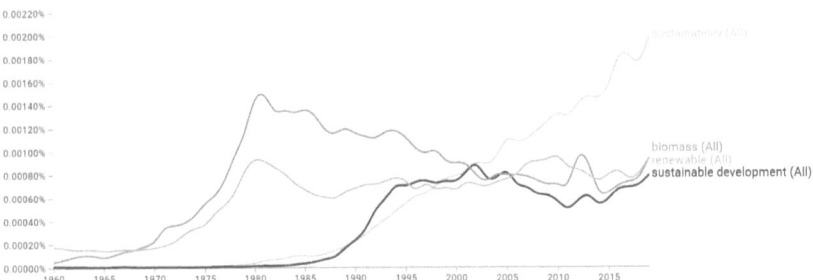

Figure 24: Popularity of terms: "sustainability," "sustainable development," "biomass," and "renewable" in the English-speaking books according to Google Ngram 1960–2019 (smoothing = 0)[30]

---

30 As with all Google Ngrams in this books, I advise caution against overstating the value of the results in absolute terms. There is plethora of variables that has to be taken into account, especially the evolving Google's book database. However, Google Ngrams are useful for analysing the moment of emergence of a given term, as well as for evaluating its relative weight compared to sister terms.

The problem is that virtually all the works on the history of sustainability assume that the concept can be used to describe the practices and ideas that had existed before the name "sustainability" appeared. In some circumstances, this is a perfectly valid reading grid and one can, for example, think about the challenges of chemurgy, of using biomass, or even about avoiding pollution in general, as a part of humanity's never-ending preoccupation with sustainability. And yet, this reading grid, reinterpreting the past through the lenses of the present, is not free from its own limitations. It may tend to ignore the alternative frameworks, as well as fail to account for the problem why the term became popular so late. In some way, just as the extension of the concept of green chemistry may lead to the impression that green chemistry has been always around (if green chemistry = chemistry of biomass, then it had been around for centuries), the same can be said about the all-embracing sustainability.

This is not to say that the concept of sustainability and its proliferation had no influence on the development of other frameworks that appeared later or in parallel, such as the EPA's green chemistry. The famous UN Rio Earth Summit took place in 1992, exactly when green chemistry was taking shape. The sustainability buzzword was penetrating the papers not only of EPA scholars, but also of all environmentally-minded chemists identifying with the green chemistry movement or not. The problem is that sustainability remained a loosely defined concept, and it was not always clear what its relationship with chemistry should be. Is green chemistry simply sustainability applied to chemistry? Or perhaps, sustainable chemistry is something different altogether? These debates, initiated in the middle of the 1990s, continue into the 2020s. In this part, I seek to retrace them and explain how the most dangerous competitor of the green chemistry framework, sustainable chemistry, came into existence.

2.1. Industrial ecology (and the life-cycle assessment)

The notion of sustainability penetrated green chemistry through interactions with various intellectual traditions; the most prominent of them in the 1990s was the discipline known under the name of industrial ecology, which, interestingly, prefigured many contemporary discussions on the bio- and circular economy. In fact, industrial ecology has been often called the science of sustainability, or the application of scientific method to the idea of sustainability. In practice, like with green chemistry, definitions are legion. One of the short and elegant definitions explains that "Industrial ecology is an approach based upon systems engineering and ecological principles that integrates the production and consumption aspects of the design, production, use, and termination (decommissioning) of products and services in a manner that minimized environmental impact while optimizing utilization of resources, energy, and capital."[31] Elsewhere we read that

> The core of industrial ecology ... is the study of the industrial metabolism. The aim is to understand the functioning of the physical basis of our societies, the interlinkages of processes and product chain webs within the 'anthroposphere' and the exchange of materials and energy with the environment. However, the interest goes further to the question of how to develop the industrial system in a way that essential requirements of sustainability can be met.[32]

Similarly to green chemistry, the definitions of industrial ecology often slightly differ between themselves to highlight different features of the concept.[33]

The standard story places the birth of industrial ecology as a discipline apart in the late 1980s, when two concepts, "industrial metabolism" and "industrial ecosystem," were combined together.[34] While the first publications on industrial ecology started appearing in the early 1990s, the first full-fledged journal in the field, *Journal of Industrial Ecology*, was established only in 1997. The concept gained rapidly a lot of attention from politicians due to its promise to thoroughly reshape the way we think about the industry and economy along sustainability lines. The same year, the Clinton White House called it "the new paradigm" in environmental protection[35] and, a few years later, the French president Jacques Chirac prefaced an introductory book to industrial ecology expressing similar hopes.[36]

---

31 Stanley Manahan, *Industrial Ecology: Environmental Chemistry and Hazardous Waste* (Boca Raton: CRC, 1999).
32 Stefan Bringezu, "Industrial Ecology and Material Flow Analysis" in *Perspectives on Industrial Ecology*, Greenleaf Publishing, ed. Dominique Bourg and Suren Erkman (London: Routledge, 2003), 20.
33 Xiaohong Li, *Industrial Ecology and Industry Symbiosis for Environmental Sustainability: Definitions, Frameworks and Applications* (London: Palgrave Pivot, 2018).
34 Robert U. Ayres, "Industrial Metabolism" in *Technology and Environment*, ed. Jesse H. Ausubel and Hedy E. Sladovich (Washington D.C.: National Academy Press, 1989): 23–49; Robert A. Frosch and Nicholas E. Gallopoulos, "Strategies for Manufacturing," *Scientific American* 261 (1989): 144–152.
35 Paul Anastas and Joseph Breen, "Design for the environment and Green Chemistry: the heart and soul of industrial ecology," *Journal of Cleaner Production* 5 (1997): 97–102; see also *Sustainable America: A New Consensus for Prosperity, Opportunity, and a Healthy Environment for the Future* (Washington D.C.: National Science and Technology Council, 1996).
36 Dominique Bourg and Suren Erkman, ed., *Perspectives on Industrial Ecology* (New York: Greenleaf Publishing, 2003).

Systems theory, life-cycle assessment, recycling, material flow analysis, clean production, bioeconomy, circular economy, and industrial symbiosis are just a few of numerous concepts that overlap, develop from, or interact with industrial ecology to various degrees. Of course, not all of them are discussed exclusively by industrial ecologists and some of them have become associated with industrial ecology even though they had a different genealogy and developed largely in parallel to it, in particular the life-cycle assessment. The plurality of concepts may be considered a strength. It suggests different communities with different disciplinary backgrounds conducted a fruitful dialogue and exchanged ideas. However, it also makes things less transparent. Navigating through these competing and complementary frameworks becomes a challenge on its own. Furthermore, in the case of green chemistry I have shown that the terms evolve through time, which makes the task of the historian even more complex. The historian has to compare ideas in a given historical moment. The problem is that their relationship can move in different directions over a few years. This was precisely what happened between industrial ecology and green chemistry.

First of all, it is worth looking at the relative popularity of terms "green chemistry," "industrial ecology" and "life-cycle assessment" through a Google Ngram analysis (Figure 25).

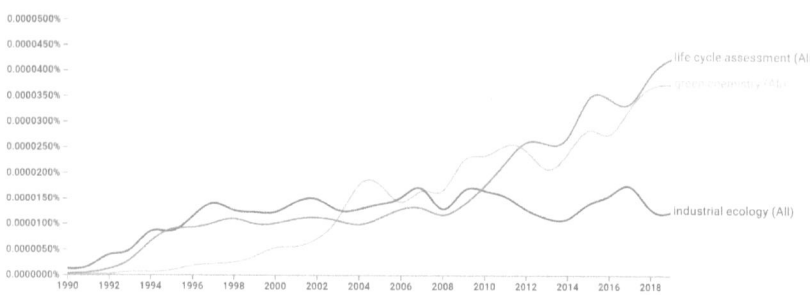

Figure 25: Popularity of "green chemistry," "life cycle assessment," and "industrial ecology" according to Google Ngram (1985–2019, smoothing = 0)

What strikes us immediately in figure 25 is the fact that these three terms are close to each other in terms of popularity. In contrast, if we added the term "sustainability," it would immediately dwarf the other ones making them look like a flat line. The three concepts are evidently used by communities of relatively similar sizes. The golden era of industrial ecology appears to stretch from 1990 to 2010. When Anastas and Warner published their book in 1998, industrial ecology was still a dominant paradigm, but it changed by the middle of the 2000s, a moment when green chemistry exploded in popularity whereas industrial ecology started losing some of its glamour (at least in comparison with other two concepts).

As for the concept of the life-cycle assessment, it is largely autonomous and has its own history and genealogy. The term was born in the 1960s, but it was not until the late 1980s when it became formalized.[37] The *Life-Cycle Assessment Journal* was established back in 1996, and the concept was also abundantly applied in the journals such as *Environmental Science & Technology*. From the very beginning, it had very close ties to industrial ecology. The terms "life cycle assessment," "life cycle analysis," "LCA," and similar have been mentioned in the *Journal of Industrial Ecology* almost 1500 times between 1997 and 2020. For comparison, they appeared only around 500 times in the *Green Chemistry* journal between 1999 and 2020. However, almost half of all the results in *Green Chemistry* came from the years 2016–2020, which suggests that there is a shift underway. At the same time, the term "industrial ecology" was featured more than 300 times in the *Life-Cycle Assessment Journal*, whereas "green chemistry" only around 30!

The debates about the relation between the LCA (and LCM – life cycle management) and industrial ecology galvanize scholars on both sides.[38] What matters, however, is that the LCA played an important role in both industrial ecology and green chemistry. What is then the difference between these two? Thomas Graedel, professor of industrial ecology, explained it in the *Green Chemistry* journal in 1999 in the following way:

> A characteristic that distinguishes industrial ecology from many other environmentally related topics is its breadth in space and time. In the case of a product, for example, it looks not just at the manufacturing stage, but at all life stages—from birth to death or, ideally, from birth to reincarnation. Such breadth has not typically been a characteristic thus far of green chemistry approaches, which focus on minimizing or avoiding residues of all kinds, especially hazardous ones, arising from the manufacturing process. This approach is laudable but unnecessarily narrow, failing to capture many environmental aspects of chemical manufacture, use and disposition. Hence, it is appropriate to expand

---

37 Jeroen B. Guinée, Reinout Heijungs, Gjalt Huppes, et al. "Life Cycle Assessment: Past, Present, and Future," *Environmental Science & Technology* 45 (2011): 90–96.

38 John R. Ehrenfeld, "Industrial Ecology and LCM: Chicken and Egg?," *The International Journal of Life Cycle Assessment* 8 (2003): 59–60.

the concept of green chemistry so as to place its approach squarely within the industrial ecology context.[39]

How should green chemists achieve this goal? Much of Graedel's article is about the life cycle assessment of chemical products. For Graedel, the LCA is what industrial ecology brings to the table when it comes to green chemistry; it is a way to improve its methods and make it more relevant. Graedel's call to put green chemistry in the industrial ecology context (a plea he reiterated twice in the article) must itself be understood through the relative position of both 'disciplines.' In 1999, green chemistry was still a small and largely unknown concept, whereas industrial ecology was in the middle of rapid expansion. It is not surprising that Graedel wanted to rally chemists to his cause and make them a part of the movement he advocated in favour of.

In 2001, he published another article, "Green chemistry as systems science," in which he writes that "Green chemistry does not operate as an isolated subsystem, but within higher levels of corporation and society. From an environmental standpoint, the ideal focus is to achieve optimum performance across the system, not at a single systems level." He later presents four levels at which green chemistry should operate. Firstly, 1) product and process levels, which are the traditional focus of green chemistry. Then, 2) the corporate level ("The environmental performance of a corporation comprises more than the greenness of its molecules. It is also related to the ways in which processes and facilities are sited, how they are developed (equipment design, building design, materials acquisition), and how they are treated upon obsolescence") at which the LCA should be incorporated. Next, 3) the infrastructure level which is about the sustainable use of feedstocks. And finally, 4) the societal level, which concerns the social acceptability of certain choices. Graedel goes as far as to suggest that "Green chemical practice must thus include using no more water than some agreed-upon allocated share."[40] In his view, green chemistry scholars should think about how to optimize all these different levels, not just the first one.

However, these were not just industrial ecology leaders that wanted to absorb green chemistry and recontextualize it within the framework of their discipline. On the contrary, already back in 1997 Anastas and his director Breen published an article "Design for the environment and Green Chemistry: the heart and soul

---

39 Thomas Graedel, "Green chemistry in an industrial ecology context," *Green Chemistry* 1 (1999): G126–G128.
40 Thomas Graedel, "Green chemistry as systems science," *Pure and Applied Chemistry* 73 (2001): 1243–1246.

of industrial ecology."[41] The paper argues that it is important "to unify, at least in understanding, the ways in which Green Chemistry and DfE [Design for Environment] work to complement, and in many cases to become, an integral part of industrial ecology and sustainable technology." The paper itself is mostly the presentation of what the benign-by-design philosophy consisted of, but the intention reiterated many times was to show how close the aspirations of green chemistry and industrial ecology were. The paper concludes by stating that "DfE and Green Chemistry are central to the development of the industrial ecology, i.e. systems approach to resolving [the] issue [of environment protection and chemical regulations]." The underlying argument was that by following the principles of green chemistry and industrial ecology, the companies would not have to worry about strict legal regulations concerning toxicity; it was a pretty typical advertising strategy from the early EPA's green chemistry promoters.

Breen was particularly fond of industrial ecology and he might have pushed green chemistry more in this direction if it was not for his sudden death in 1999. However, Anastas himself referred to industrial ecology many times in his papers from the late 1990s and early 2000s, for example in his article with the telling title "Life cycle assessment and green chemistry: the yin and yang of industrial ecology" published together with Rebecca Lankey in *Green Chemistry* in 2000.[42] This short introductory paper did not attempt to recast green chemistry as some sort of all-embracing systems science like Graedel did, but to simply advocate the use of the LCA in green chemistry. For Anastas and many others, the three concepts from the paper's title were interwoven. I mention it also to reiterate that the frontiers of green chemistry were porous and its founding fathers, be it Anastas, or Warner, or Breen, were eager to contextualize it within the more successful and better recognized frameworks (over the period) such as sustainability or industrial ecology, to promote their own ideas. At the same time, it begs once more questions about what green chemistry actually was meant to be, and why the tools such as the LCA were sometimes placed at the heart of the green chemistry project, and sometimes barely mentioned in the discipline's programmatic papers.

Overall, while the LCA became, not immediately but over the course of years, the goal to strive for in green chemistry, the term industrial ecology did not meet

---

41 Paul Anastas and Joseph Breen, "Design for the environment and Green Chemistry: the heart and soul of industrial ecology," *Journal of Cleaner Production* 5 (1997): 97–102.
42 Paul Anastas and Roger Lankey, "Life cycle assessment and green chemistry: the yin and yang of industrial ecology," *Green Chemistry* 2 (2000): 289–295.

the same fate. A full-text search in *Green Chemistry* reveals around 25 articles mentioning "industrial ecology" between 1999 and 2020, a drop in the ocean. Conversely, the term "green chemistry" has been mentioned in more than 2150 publications of the *Journal of Industrial Ecology* between 1997 and 2020. These results may be read in very different ways. They show that industrial ecology scholars are more open to the use of other concepts, such as green chemistry, because of the inherently integrative and systemic approach of their discipline. But they also show that green chemistry scholars themselves were much less interested in the system thinking and did not follow Graedel's ideas on how to extend green chemistry objectives onto higher-level subsystems.

And yet, this is not entirely true. The systemic approach appears more and more frequently in green chemistry-related publications, but it is associated with a different concept: sustainable chemistry.

2.2. 1998 OECD Sustainable Chemistry workshop

In 1998, OECD launched its initiative on "sustainable chemistry" with an international workshop in Venice. While the term itself had circulated before, it is in 1998 when it really took off and its formalisation began. The fruit of the workshop was a 1999 report that laid down the foundations of the concept. In this publication, there is the following definition of the new term:

> Within the broad framework of sustainable development, one should strive to maximise resource efficiency through activities such as energy and non-renewable resource conservation, risk minimisation, pollution prevention, minimisation of waste at all stages of a product's life-cycle, and the development of products that are durable and can be re-used and recycled. Sustainable Chemistry strives to accomplish these ends through the design, manufacture and use of efficient and effective, more environmentally benign chemical products and processes.[43]

There are many elements here that we have seen before in the discussions on green chemistry, such as pollution prevention and waste minimization, along with some ideas that were usually more associated with industrial ecology (at least in the 1990s): life-cycle and recycling. However, these additional elements should not be overestimated as throughout the report the concepts of green and sustainable chemistry were not strictly separated. We read, for example, in one of the report's papers that: "Sustainable Chemistry is the design of chemical

---

43 "Executive Summary," *Proceedings of the OECD workshop on sustainable chemistry, Series on Risk Management OECD* (Paris: OECD, 1999): 13–14, 13.

products and processes that reduce or eliminate the use and generation of hazardous substances," which is more or less the early definition of the EPA's green chemistry (benign from design).[44] The author, Joseph Carra, was effectively an EPA employee. The article explains that the major technical focus areas of sustainable chemistry should be: the use of alternative syntheses, the use of alternative solvents and reaction conditions, and the design of safer chemicals. Moreover, Carra states at one point that green and sustainable are just two names for the same concept, one used in the US, the other adopted by OECD, therefore there is no theoretical difference between the two.

Out of 25 papers (and posters) presented in the 1998 OECD workshop, five had "green chemistry" in their titles, and eleven papers had "sustainable chemistry." All five green chemistry papers were written by American scholars, including Paul Anastas, John Warner, and Joseph Breen, who put a lot of effort to describe the activities of, among others, the EPA and the Green Chemistry Institute. As for the papers using the term "sustainable chemistry," five were written by researchers from Japan, five from Germany and Austria, and one was the previously mentioned article by Joseph Carra. With the exception of the last paper, the terminological difference appears to run much more along national than conceptual lines.

One of the Japanese papers, "The Japanese approach to Sustainable Chemistry," offers perhaps the most refined definition of the concept adopted by the Japanese Ministry of International Trade and Industry. The Japanese definition of sustainable chemistry covers: "Sciences and technologies aiming to reduce the adverse effects and/or increase the positive contribution to human health and the environment by chemicals at every stage of their life cycle, i.e. raw materials, production, utilization, disposal and recycling of chemical substances and products"[45] The Japanese sustainable chemistry comprised 6 categories: Green Feedstock (e.g. renewable resources), Green Process, (e.g. simple chemical processes), Green Product (e.g. non-persistent chemicals), Green Treatment/Remediation (e.g. zero emission), Green Recycling (e.g. material recycling, chemical recycling), and Green Infrastructure (e.g. LCA, Assessment Methodology).

---

44 Joseph Carra, "International Diffusion of Sustainable Chemistry," *Proceedings of the OECD workshop on sustainable chemistry, Series on Risk Management OECD* (Paris: OECD, 1999): 47–50, 48.

45 Hisao Ida, "The Japanese approach to Sustainable Chemistry," *Proceedings of the OECD workshop on sustainable chemistry, Series on Risk Management OECD* (Paris: OECD, 1999): 73–78, 73.

Again, this is larger than the EPA's green chemistry, but the authors were content with using the green label as well and clearly indicated that the terms "green" and "sustainable" were often used interchangeably by the Japanese themselves.

Reading these various contributions may give the impression that green chemistry and sustainable chemistry were overall just two names for more or less the same thing. The nuances, if there were any, were marginal and resulted from some differences in national vocabularies. Sustainable chemistry was perhaps slightly larger but, after all, one could assume that the exact names do not matter that much if there is a common goal. The problem is that, in reality, the names and genealogies do matter and the 1999 report does not account for the debates concerning both terms that took place during the 1998 workshop.

The Canadian scholar Ian Brindle, described in one of the 1999 editorials in the *Green Chemistry* journal a discussion that took place during the OECD's meeting on sustainable chemistry:

> Problems arose from the difficulties that the delegates had trying to incorporate the perspective of the Brundtland Report on sustainability into the already established notion of green chemistry. The idea of sustainable chemistry creates a new set of problems. Green chemistry is, in my view, remains a better descriptor than sustainable chemistry. Extraction of non-renewable resources is, by its nature, not sustainable and so the notion of doing sustainable things with an unsustainable end sounds perverse.[46]

What was Brindle referring to? I have already mentioned in chapter 2 that he discussed the application of green chemistry principles to mining-related operations. He believed this type of activities could be 'greened,' but since this chemistry concerned non-renewable feedstock, it should not be called sustainable. The key takeaway from Brindle's article is that any consensus over the matter was illusory. Green chemistry had its advocates that considered the concept of sustainable chemistry misleading, and vice versa, some sustainable chemistry partisans opposed the use of the green terminology for the reasons I will make clear in a while.

The debate over the choice of wording was also divisive for international organisations such as the International Union of Pure and Applied Chemistry. One of IUPAC's papers on more environmentally friendly chemical syntheses, co-authored by Anastas and based on the work of the IUPAC Working Party on Synthetic Pathways and Processes in Green Chemistry, explained the problem in the following terms in 2000:

---

46 Ian D. Brindle, "Green chemistry—a Canadian perspective," *Green Chemistry* 1 (1999): 155–157.

The terminology "green chemistry" or "sustainable chemistry" is the subject of debate. The expressions are intended to convey the same or very similar meanings, but each has its supporters and detractors, since "green" is vividly evocative but may assume an unintended political connotation, whereas "sustainable" can be paraphrased as "chemistry for a sustainable environment," and may be perceived as a less focused and less incisive description of the discipline. Other terms have been proposed, such as "chemistry for the environment" but this juxtaposition of keywords already embraces many diversified fields involving the environment, and does not capture the economic and social implications of sustainability. The Working Party decided to adopt the term green chemistry for the purpose of this overview. This decision does not imply official IUPAC endorsement for the choice. In fact, the IUPAC Committee on Chemistry and Industry (COCI) favors, and will continue to use sustainable chemistry to describe the discipline.[47]

To sum up, even in a single organisation different vocabularies competed one with another. While the Americans, supported by the EPA and the Green Chemistry Institute, tended to promote the concept of "green chemistry" and use this term in their own publications, it was Germans who led the charge against it and advocated in favour of the term "sustainable chemistry."

2.3. German connection

Otto Hutzinger from the University of Bayreuth explained the German difficulties with the word "green" in 1999 in the following way:

> Whilst most scientists use English as the lingua franca of science, some argue that cultural-sociological factors giving different meanings to terms and words should also be considered. "Green," for instance, in a cultural context simply means something different in countries such as the U.S. and Germany. ...
> In Germany, the term "green" in many elicits political-sociological feelings – for instance, fear of production shutdowns in chemical industry – and general opposition to genetic engineering projects and atomic energy by members of the Green Party.
> There has been strong opposition against the term "green chemistry" by IUPAC, OECD, CEFIC and GDCh headed by the German speaking members. Further, "green" was abandoned by the European Commission in the section of the 5th Framework program, likewise headed by the German speaking members.
> The underlying meaning of the terms "Green Chemistry" and "Sustainable Chemistry" is different. Sustainable Chemistry is the maintenance and continuation of an ecologically sound development whereas Green Chemistry focuses on the design, manufacture, and use of chemicals and chemical processes that have little or no pollution potential or

---

47 Pietro Tundo, Paul Anastas, David StC. Black et al., "Synthetic pathways and processes in green chemistry. Introductory overview," *Pure and Applied Chemistry* 72 (2000): 1207–1228, 1207.

environmental risk and are both economically and technologically feasible. The principles of green chemistry can be applied to all areas of chemistry including synthesis, catalysis, reaction conditions, separations, analysis, and monitoring. Thus, the terms cannot easily be exchanged.[48]

Hutzinger's commentary is very informative. Firstly, he explains that the opposition to the term "green" in Germany was driven by the political connotation of the word, which was associated with the activism of the German green party (Bündnis 90/Die Grünen) traditionally strongly critical of the chemical (and nuclear) industry. The terminology and the metaphors of green chemistry may have been too close to the discredited sanfte Chemie and elicited fears in the German industrial establishment. Secondly, Hutzinger explains that the representatives of German industry occupying prominent positions in the international bodies (OECD, IUPAC, etc.) actively fought against their Anglo-Saxon colleagues to keep the "green vocabulary" from spreading. In fact, Hutzinger mentions later the debates concerning the naming of the European Green Chemistry Award that was renamed to the European Green and Sustainable Chemistry Award subsequently, following the political pressure coming from the German chemists.

At the same time, Hutzinger concedes that green and sustainable chemistry are not identical. In his view, the concept of green chemistry is much more meaningful and narrower. This remark should not be not surprising. By 1999, the theoretical backbone of the green chemistry concept was already solidly established by the EPA. This was not the case of the somewhat elusive idea of sustainable chemistry that had been tentatively defined in different contexts but remained very intuitive. Finally, it is worth mentioning that Hutzinger's article was published in the journal *Environmental Science and Pollution Research*. Environmental science was still the framework of choice for these debates.

Overall, what matters is that the Germans preferred the term "nachhaltige Chemie" [sustainable chemistry] and not "grüne Chemie" [green chemistry]. In fact, the German federal government used the concept of nachhaltige Chemie already in its internal documents in 1997 to describe big lines along which German industry should develop.[49] What matters is that Germany was an outlier, as the tendencies in other linguistic zones developed in a different direction. Figures 26, 27, and 28 present Google Ngram plots illustrating the scale of the

---

48 Otto Hutzinger, "The Greening of Chemistry- Is It Sustainable?" *Environmental Science and Pollution Research* 6 (1999): 123.
49 Eherhard Weise, Henning Friege, Karl Otto Henseling, et al., "Wie die Chemie "grün" wurde," *Nachrichten aus Chemie, Technik und Laboratorium* 47 (1999): 914–918, 918.

discrepancy between the terminology in three languages: English, French, and German.

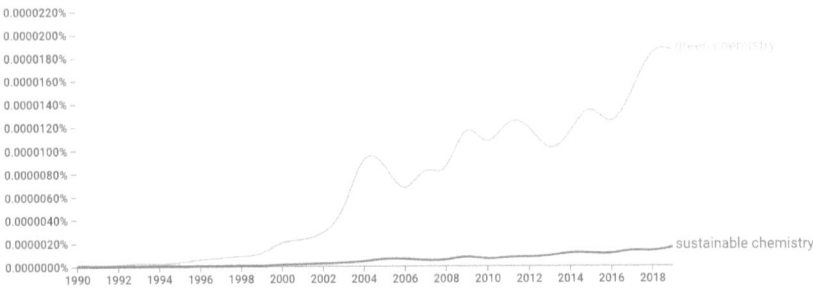

Figure 26: The use of the terms "sustainable chemistry" and "green chemistry" according to Google Ngram (English literature, 1990–2019, smoothing = 0)[50]

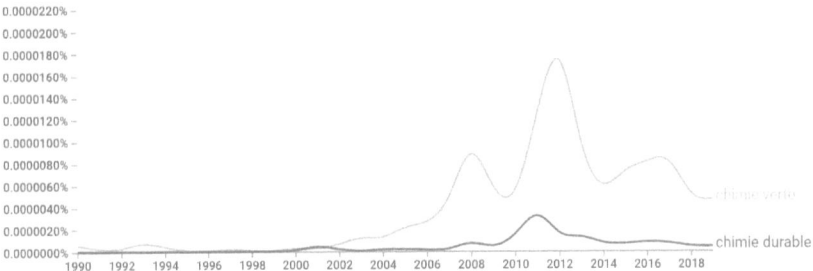

Figure 27: The use of the terms "chimie durable" [sustainable chemistry] and "chimie verte" [green chemistry] according to Google Ngram (French literature, 1990–2019, smoothing = 0)

---

50 These Ngrams are case-sensitive, unlike many others presented in this book. This is due to the technical glitch preventing the French literature from analysis otherwise.

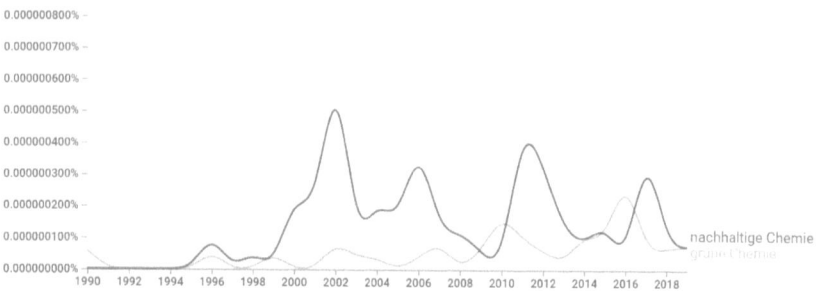

Figure 28: The use of the terms "nachhaltige Chemie" [sustainable chemistry] and "grüne Chemie" [green chemistry] according to Google Ngram (German literature, 1990–2019, smoothing = 0)

In the French- and English-speaking scientific literature, the term sustainable chemistry is dwarfed by green chemistry. Sustainable chemistry remained marginal even in France, despite the fact that the French National Scientific Research Centre (CNRS) and the French National Research Agency (ANR), favoured the term "chimie durable" to fund projects that would have probably been referred to as green chemistry in the US and the UK (especially on biomass).[51] On the other hand, the German-language scientific publications clearly favour nachhaltige Chemie over grüne Chemie. This observation should be nuanced though. In practice, the Germans were so reluctant to talk about grüne Chemie for political reasons that they used the English term "green chemistry" even in the articles and books written in German. Even today, when the opposition to the word "grüne" appears to have waned, there are German-speaking articles using the English name instead.[52] If we analyse the German books and combine the terms "grüne Chemie" and "green chemistry" into a single category, the advantage of sustainable chemistry is less pronounced, although this term is still much more popular in German than in English and French (Figure 7). As a side-note: if we look for the term "green chemistry" in English in the French Google Ngram book corpus, we get almost no result.

---

51 Estelle Garnier, "Une approche socio-économique de l'orientation des projets de recherche en chimie doublement verte" (PhD diss., University Reims Champagne-Ardenne, 2012), 248.
52 Michael Linkwitz and Ingo Eilks, "Green Chemistry in der Schule," *Chemie in unserer Zeit* 53 (2019): 412–420.

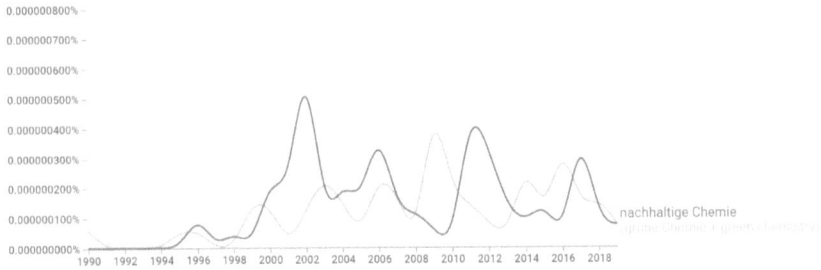

Figure 29: The use of the terms "sustainable chemistry" (nachhaltige Chemie) and the combined category of "green chemistry" + "grüne Chemie" in German-speaking books according to Google Ngram (1990–2019, smoothing = 0)

How to explain the particularity of the German case? I have already mentioned the grudge held by German chemical industry against green political movements, but there is more to it. For example, the term Nachhaltigkeit [sustainability] was coined by the German foresters already back in the eighteenth century. The long-lasting Nachhaltigkeit tradition may have legitimized the nachhaltige Chemie wording by connecting it to a conceptual grid with which many German scholars were more familiar. Another indirect reason to reject the green chemistry vocabulary was the fact that in Germany there had been already different traditions dealing with similar challenges. The previous section explored the radical sanfte Chemie and a much more conservative Claus Christ's "Production-Integrated Environmental Protection and Waste Management in the Chemical Industry." To these two we can add the tradition of ökologische Chemie [ecological chemistry] created by Friedhelm Korte in the 1970s. The story goes that unlike Umweltchemie [environmental chemistry], which was focused on chemical processes in the environment in general, ökologische Chemie studied specifically the influence of pollution on the environment, with a special interest in ecotoxicology.[53]

What matters for now is that when the American green chemistry arrived in Germany, it was merely another concept trying to frame the debates that had been going on for decades. The German chemists did not necessarily see it as something particularly revolutionary and thus may have been more reluctant to embrace it, especially because it lacked some more overarching features of native

---

53   Stefan Böschen, Dieter Lenoir, and Martin Scheringer, "Sustainable chemistry: starting points and prospects," *Naturwissenschaften* 90 (2003): 93–102.

concepts such as ökologische Chemie, or had little added value as compared with Claus Christ's conceptualisation. As a consequence, the broader, less defined and more ecumenical idea of sustainable chemistry fell on more fertile ground in the German-speaking world.

2.4. American trajectory: sustainable chemistry is green chemistry

And yet, the fact that the German scholars chose to use a different term does not imply that there must have been a meaningful difference between sustainable and green chemistries. In fact, for many years, chemists all over the world continued to indiscriminately equate the two and in a great many publications the terms have been used interchangeably, just like in the OECD's 1999 report. According to this first narrative, green chemistry is basically the same thing as sustainable chemistry.

For example, a short 2001 *Science* article published under the title "Towards Sustainable Chemistry" fully equates the two concepts and talks about the 12 principles of green *and* sustainable chemistry.[54] Similarly, the US Congress' report on the Green Chemistry Research and Development Act from 2005 states that green, benign, and sustainable chemistry are the same thing.[55] The same position can be found in many textbooks, also very recent ones. For example, an influential *Environmental Chemistry* handbook (2017, 10th edition) by Stanley E. Manahan, simply states that: "Green chemistry is sustainable chemistry" and does not develop the topic any further.[56] The 2016 book *White Biotechnology for Sustainable Chemistry*, in spite of its title, mentions the term "sustainable chemistry" only seven times in the entire book (above all in the preface), whereas the term "green chemistry" appears in it 105 times.[57] Roger Sheldon, the father of E-factor and the leading figure of the green chemistry movement wrote at one point that "One could say that sustainability is our ultimate common goal and green chemistry is the means of achieving it."[58] In 2016,

---

54 Terry Collins, "Towards Sustainable Chemistry," *Science* 291 (2001): 48–49.
55 Sherwood Boehlert, Report prepared for the Green Chemistry Research and Development Act of 2005, report No. 109-82, 2005, 5.
56 Stanley Manahan, *Environmental Chemistry* (Boca Raton: CRC Press, 2017), 453.
57 Maria Alice Coelho and Bernardo D. Ribeiro, *White Biotechnology for Sustainable Chemistry*, (Cambridge: RSC Green Chemistry, 2016).
58 Roger Sheldon, "The E Factor: fifteen years on," *Green Chemistry* 9 (2007): 1273–1283.

Anastas simply said: "if Sustainability is the goal, Green Chemistry will show the way!"[59]

In fact, Anastas and other EPA-connected scientists insisted on this interpretation from very early on. For example, in a 2002 volume co-edited by Anastas entitled *Advancing Sustainability through Green Chemistry and Engineering*, the introductory chapter discusses the broad challenges of sustainability such as climate change, energy production, food production, resource depletion, and toxic pollution in the environment, and argues that the application of principles of green chemistry and green engineering can help to address them.[60]

The idea that green chemistry is essential in making the industry more sustainable was advocated in the 2006 report *Sustainability in the Chemical Industry, Grand Challenges and Research Needs* of the American National Research Council. The report identified eight grand challenges that the chemical industry had to face in the following 100 years (Table 28).[61]

Table 28: Grand Challenges of Sustainability

1. Green and Sustainable Chemistry and Engineering ("Discover ways to carry out fundamentally new chemical transformations utilizing green and sustainable chemistry and engineering, based on the ultimate premise that it is better to prevent waste than to clean it up after it is formed. Over the next twenty years this will involve replacing harmful solvents or improving catalytic selectivity and efficiency in chemical reactions that also provides cost savings. This area will grow in importance as fossil fuels are phased out of use and alternative and innovative approaches are required.")
2. Life Cycle Analysis
3. Toxicology
4. Renewable Chemical Feedstocks
5. Renewable Fuels
6. Energy Intensity of Chemical Processing
7. Separation, Sequestration, and Utilization of Carbon Dioxide
8. Sustainability Education

The development of green *and* sustainable chemistry was the first grand challenge of sustainability for the twenty-first century according to the American

---

59 Paul Anastas, Buxing Han, Walter Leitner, Walter Leitner and Martyn Poliakoff, "'Happy silver anniversary:' Green Chemistry at 25," *Green Chemistry* 18 (2016): 12–13, 12.
60 Rebecca L. Lankey and Paul Anastas, ed., *Advancing Sustainability through Green Chemistry and Engineering* (Washington D.C.: American Chemical Society, 2002).
61 National Research Council, *Sustainability in the Chemical Industry, Grand Challenges and Research Needs* (Washington D.C.: The National Academies Press, 2006).

National Research Council, but both terms were understood above all as the EPA's green chemistry. Anastas and Warner's 1998 book (both of whom contributed to the report) was in fact cited abundantly in the text.

All of this is to say that Anastas, Warner, EPA collaborators, and the broader green chemistry community in the United States were supportive and interested in the sustainability-related challenges and believed that green chemistry is an essential concept to address them. Green chemistry was for them what chemistry could bring to the table when it comes to sustainability.

The association of the two concepts was perpetuated in the public understanding by Wikipedia. The Wikipedia article on "green chemistry" was modified on 23 November 2007 when a seemingly inconspicuous edit changed the beginning of the article from "green chemistry is" to "green chemistry, also called sustainable chemistry, is."[62] Not only has not this definition been challenged between 2007 and 2022, but over the same period there has been a link in Wikipedia redirecting from "sustainable chemistry" to "green chemistry" effectively presenting both approaches as entirely equivalent to a non-professional public.

Finally, the similarity of green and sustainable chemistry can be confirmed by a scientometric analysis. While the number of papers tagged as green chemistry is of a different order of magnitude than the number of articles labelled as sustainable chemistry (Scopus indexes around ten times more articles in green than sustainable chemistry), the overall trend evolved along the same lines (Figure 30).

---

62 The modification of the Green Chemistry article in Wikipedia from 23 November 2007, accessed 22/02/2022, https://en.wikipedia.org/w/index.php?title=Green_chemistry&diff=173285727&oldid=173250886.

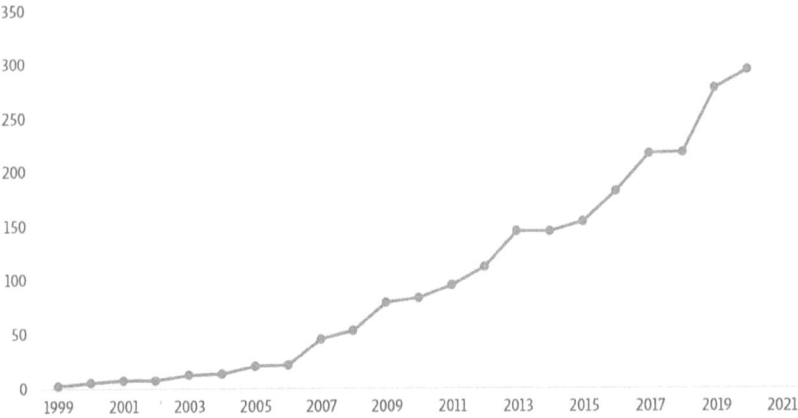

Figure 30: Number of papers published yearly with "sustainable chemistry" in their titles, abstracts, or keywords, according to Scopus.

The Scopus keyword analysis for the years 1999–2020 indicates that catalysis, biomass, carbon dioxide, synthesis, and ionic liquids were the five most popular keywords appearing in sustainable chemistry papers (after cleaning the dataset). The high place of biomass is remarkable, but the other keywords are fairly in line with the mainstream green chemistry, even if carbon dioxide is somewhat more ambiguous, a problem to which I immediately return. The most cited article tagged with the "sustainable chemistry" keyword in Scopus is about catalysis,[63] and the second one is on biomass and catalysis;[64] topics that could be easily labelled as green chemistry as well. Their noteworthy feature is that they both appeared in the pages of *Angewandte Chemie*. It shows again that in the German publications there was a certain reluctance to embrace the green chemistry label.

Is there anything that would warrant any different understanding of sustainable chemistry according to our Scopus results? Yes, the third most-cited article,

---

63 Yong Wang, Xinchen Wang, and Markus Antonietti, "Polymeric graphitic carbon nitride as a heterogeneous organocatalyst: From photochemistry to multipurpose catalysis to sustainable chemistry," *Angewandte Chemie – International Edition* 51 (2012): 68–89.

64 Juben N Chheda, George W Huber, and James A. Dumesic, "Liquid-phase catalytic processing of biomass-derived oxygenated hydrocarbons to fuels and chemicals," *Angewandte Chemie – International Edition* 46 (2007): 7164–7183.

published in *ChemSusChem*, concerns CO_2 capture.[65] CO_2 is one of the keywords often associated with green chemistry because supercritical CO_2 can be used as a solvent, but traditionally construed green chemistry would not deal with the topics such as CO_2 capture (which is, *de facto*, a type of pollution remediation). Is this article some isolated exception artificially stretching the concept of green/sustainable chemistry to cover the problems explored by other disciplines? Not at all, it indicates a new distinct tradition of sustainable chemistry slowly affirming its autonomy.

2.5. Going beyond green: early years of sustainable chemistry (1998–2011)

Already in the early 2000s, there were occasionally definitions of sustainable chemistry that came from different traditions and argued for the concept's autonomy in its relation to green chemistry. Unsurprisingly, one of the first full-fledged and explicit definitions of such sustainable chemistry was formulated by a group of German scholars led by the scientist-turned-sociologist Stefan Böschen in 2003:

> Here we propose a working definition of sustainable chemistry which comprises two elements: (1) at the level of energy and material flows, the aim of sustainable chemistry is to optimise processes and products with respect to resource consumption, waste generation and properties of products (low toxicity, efficiency, degradability etc.); (2) with respect to the interaction with non-scientific actors, the aim is to reconcile and coordinate scientific objectives with non-scientific demands, preferences and values; to this end, it is necessary to make transparent the assumptions as well as objectives and implications of scientific research[66]

Böschen considers in his article the EPA's green chemistry, understood mostly as green synthesis, to be one of the tools for achieving the goal of a much more ambitious project of sustainable chemistry. This project would include concepts and practices such as green synthesis, renewability of feedstocks, and life-cycle assessment, but its main remarkable feature would be above all the openness to social dialogue; this sustainable chemistry would be actively engaged in solving social problems. Böschen makes part of a different trajectory than American green chemists and cites mostly German papers, including the works of the

---

65 Sunho Choi, Jeffrey H. Drese, and Christopher W. Jones, "Adsorbent materials for carbon dioxide capture from large anthropogenic point sources," *ChemSusChem* 2 (2009): 796–854.

66 Böschen, Lenoir, Scheringer, "Sustainable chemistry: starting points and prospects," 97.

father of sanfte Chemie, Arnim von Gleich, and the father of ecological chemistry, Friedhelm Korte.

Nevertheless, Böschen's paper does not radically differ from what was discussed over the same period in the US. The 2006 sustainability report of the National Research Council also discussed LCA, renewability, toxicology, and public outreach, even if, effectively, the social dialogue was somewhat less put forward in the American publications. The primary difference lies, however, in the delineation of the frontiers. For the American green chemists in the middle of the 2000s, there was green chemistry, and there were these other important tools or ideas for achieving sustainability. For Böschen, they all were part of a single grand project advanced under the label of sustainable chemistry.

As interesting as these subtleties are, none of this warrants yet the creation of a new discipline. German scholars simply participated in a very vivid discussion led on a global level about making chemistry more environmentally friendly, and their ideas were ones of many. It means that the 2003 paper should not be seen as some sort of foundational paper for the field of sustainable chemistry as it did not explicitly try to construct a new framework.

One year later, in 2004, an international congress entitled "Sustainable Chemistry – Integrated Management of Chemicals, Products and Processes" was organised by the OECD and by the German Federal Ministry of the Environment, Nature Conservation and Nuclear Safety in Dessau, Germany. It was a follow-up to another conference held in 2002 in Johannesburg, where the topic of sustainable chemicals and processes had been already discussed.[67] None of the papers presented in the 2004 congress tried to frame sustainable chemistry as a separate field of study, but the scope of the discussions clearly indicated that it was more than just green chemistry (although Paul Anastas was the congress's keynote speaker) or at least that the concepts were not entirely equivalent. For example, some papers brought up problems with the regulation of toxic chemicals, consumer information, and of chemical leasing. Chemical leasing is certainly a topic that deserves to be discussed in the context of making the chemical industry more sustainable, but it goes far beyond what is usually considered green chemistry.

Two years after the congress, in 2006, a German-speaking book *Nachhaltige Chemie. Erfahrungen und Perspektiven* [*Sustainable Chemistry: Experiences and Perspectives*] came out. This publication was a collaborative work in which

---

67 Klaus Günter Steinhäuser, Steffi Richtera and Jutta Penninga, "Sustainable Chemistry in Dessau – a workshop report," *Green Chemistry* 6 (2004): G41–G43.

different chapters were written by different authors. Its overall ambition was to connect a range of topics at the frontier of sustainability and chemistry.[68] Again, its authors did not argue that sustainable chemistry is somehow different from the green one. On the contrary, the first chapter, written in English by a British scholar Jeff Hardy, "The Politics and Practice of Sustainable Chemistry in the UK," considers both concepts to be synonymous. The German authors in the following chapters are also relatively liberal when it comes to definitions and frequently mention the 12 principles and the EPA's contributions to the field as an example of sustainable chemistry. And yet, throughout the entire volume there is a distinct feeling that green chemistry is only a part of a broader idea; that it is only a means to a more ambitious end, which is to make chemistry more sustainable on many different levels. The scope of the 2006 book extends far beyond green synthesis and covers themes such as biomimetics, the use of microorganisms for manufacturing (white biotechnology), and non-scientific managerial concepts such as the already mentioned chemical leasing.

A similar philosophy appears to have led to the establishment of *ChemSusChem* in 2008, arguably the first journal in sustainable chemistry properly speaking. The journal covers topics traditionally associated with green chemistry (e.g. catalysis for pollution prevention), topics whose attachment to green chemistry used to be somewhat ambiguous (e.g. biomass, biofuels, renewability), and topics that are clearly outside green chemistry's scope (e.g. photovoltaics, CO_2 capture). One of the founders of *ChemSusChem* explained that the establishment of the journal had been preceded by discussions "about the necessity for a journal focusing on chemical aspects of sustainability. Sustainability is more than being green!"[69] In other words, there was a vague feeling in the late 2000s that something had been left out by practitioners of green chemistry so far. This positioning can be explained by the fact that *ChemSusChem* was owned by the European Union of Chemical Societies, and that the editorial team was based in Germany and associated with *Angewandte Chemie*'s family of journals.

In the late 2000s and early 2010s, the frontier between green and sustainable chemistries has been put forward more and more often. For example, in the book *Sustainable Industrial Processes* from 2009, one of the first chapters entitled "Green versus Sustainable Chemistry" mentions the distinction between the two chemistries made by Hutzinger in 1999 and then adds:

---

68 Michael Angrick, Klaus Kümmerer and Lothar Meinzer, *Nachhaltige Chemie. Erfahrungen und Perspektiven* (Marburg: Metropolis, 2006).
69 David J. Smith, "A Fresh Look at Sustainable Chemistry," *ChemSusChem* 14 (2021): 5–9.

sustainable chemistry is not synonymous with green chemistry, ... [it has] a lower impact on environment and human health, but goes beyond the latter ..., seeing chemistry as part of an integrated vision where chemistry, sustainability and innovation are three key components for the future of our society.[70]

In 2011, a book entitled simply *Sustainable Chemistry* was published as a follow-up to the "First International Conference on Sustainable Chemistry" co-organized by the Wessex Institute of Technology and the University of Antwerp.[71] The conference included six sections: eco-efficiency, smart processing technology for sustainability, improvements in catalysis, multifunctional materials, bio-based materials, and environmental health issues. While some topics, especially catalysis but also, to some extent, bio-based materials, overlapped with traditionally construed green chemistry (as of 2011 at least), the others steered again in a very different direction. Perhaps, the most telling example was the introductory chapter co-authored by Genserik Reniers, an economist and one of the volume's editors. The chapter "How to enhance sustainable chemistry in a non-technological way?" explains that:

> To achieve sustainable industrial chemical processes and products, companies, research centers and academia tend to focus mainly on technological solutions such as cleantech, green technology, process intensification, new catalysts, new membranes, ecofining, etc. However, non-technological approaches are essential as well to succeed in adequate sustainable chemistry. Cluster management, sustainable supply chain management, chemical leasing, integrated management systems and business models, societal expectations, etc. are all important non-technological aspects of sustainable chemistry.[72]

The focus on the non-technological aspects of the chemical enterprise became a defining element of this new line of thinking.

2.6. Defining sustainable chemistry (2011–2017)

In the 2010s, three important journals propelled forward the development of the concept of sustainable chemistry: *ACS Sustainable Chemistry and Engineering*, *Sustainable Chemistry and Pharmacy*, and *Current Opinion in Green and Sustainable Chemistry*. None of them was fully devoted to the new concept, but they all

---

70 Gabriele Centi and Siglinda Perathoner, "From Green to Sustainable Industrial Chemistry," in *Sustainable Industrial Processes*, ed. Fabrizio Cavani, Gabriele Centi, and Siglinda Perathoner (Weinheim: Wiley-VCH, 2009), 1–69, 5.
71 Genserik Reniers and Carlos A. Brebbia, *Sustainable Chemistry* (Antwerp: WIT Press, 2011).
72 Reniers and Brebbia, *Sustainable Chemistry*, 3.

used the new term abundantly to describe their content. In practice, however, in spite of the similarity of their names, they stem from two different traditions. The first journal, *ACS Sustainable Chemistry and Engineering*, was established under the auspices of the American Chemical Society in 2013 and is deeply grounded in the American research culture, whereas the other two are rooted in German scholarship. It means that the way they approach sustainable chemistry also differs. For instance, the editorial to the first issue of *ACS Sustainable Chemistry and Engineering* explains that: "This first issue is the culmination of several years of planning within the American Chemical Society for a journal dedicated to publishing high impact research advances in the fields of green chemistry, green engineering, and sustainability."[73] In other words, the term "sustainable chemistry" is not even mentioned as something autonomous. Furthermore, the first issue included topics such as biomass, solvents, and green metrics; topics that could be found in the pages of other green chemistry journals. Can we then say that for the *ACS Sustainable Chemistry and Engineering* editors green and sustainable chemistry were fully equivalent? It certainly appears so to many authors publishing in the journal, but the analysis of the editorials throughout the years points out that there was a shift under way.

In 2020, an editorial explains that *ACS Sustainable Chemistry and Engineering* is a forum

> for our authors and readers working in areas of (I) catalysts for sustainable chemical transformations, (ii) renewable materials, (iii) electrochemistry, photochemistry, and photoelectrochemistry for energy conversion, energy storage, and synthesis of value added chemicals, (iv) benign solvents, (v) biorenewable feedstocks in fuel and chemical manufacturing, (vi) sustainable chemical synthesis, (vii) design of sustainable chemical processes, (viii) industrial ecology and sustainable chemicals management, and (ix) the use of advanced nanomaterials for sustainable chemistry/engineering applications.[74]

Here, it can be easily observed that the scope of the journals was broadened. Not only does it cover the topics traditionally outside of the scope of green chemistry (e.g. energy storage), but the connection to industrial ecology and chemicals management marks an opening towards system thinking. Still, it may be argued that this evolution can be interpreted precisely as the opening of green chemistry to novel approaches. In other words, it does not denote a shift from

---

73 David Allen, "Welcome to ACS Sustainable Chemistry & Engineering," *ACS Sustainable Chemistry and Engineering* 1 (2013): 1.
74 David Allen, Julie Carrier, Jingwen Chen et al. "The Evolution of ACS Sustainable Chemistry & Engineering," *ACS Sustainable Chemistry and Engineering* 8 (2020): 1–1.

green to sustainable chemistry, as if these were two separate ideas, but the redefinition of the frontiers of the former. After all, many scholars continue to use the terms green and sustainable interchangeably without trying to delineate the two concepts from a theoretical standpoint at all.[75]

This is, however, certainly not the case of two Germany-based journals, *Sustainable Chemistry and Pharmacy* and especially the *Current Opinion in Green and Sustainable Chemistry*, both co-created by Klaus Kümmerer, a leading German figure of environmentally friendly thinking in chemistry and probably the most active promoter of sustainable chemistry as a separate idea today, who was also once involved in the sanfte Chemie group. The first issue of the *Current Opinion* from 2016 was a follow-up to the "1st Green and Sustainable Chemistry Conference" hosted the same year in Berlin (it is worth mentioning that the 1st Conference of Sustainable Chemistry that took place in Antwerp in 2011 was not continued). At first sight, it appears that the journal follows the same philosophy as *ACS Sustainable Chemistry and Engineering*; that it considers green and sustainable chemistry to be just one single discipline. Without a doubt, this was the case for many authors publishing in its pages. Nonetheless, the analysis of the content of the journal clearly indicates that its editorial board has a distinct and original approach. This novelty is particularly striking in the papers published in the journal's pages after the second Green and Sustainable Chemistry conference held in 2018 (table 29).[76]

---

75 Mark Burgman, Mike Tennant, Nikolaos Voulvoulis et al., "Facilitating the transition to sustainable green chemistry," *Current Opinion in Green and Sustainable Chemistry* 13 (2018): 130–136; Avtar Matheru and Pascale Champagne, "Brown to green and sustainable chemistry," *Current Opinion in Green and Sustainable Chemistry* 2 (2016): iii-iv.
76 *Current Opinion in Green and Sustainable Chemistry* 9 (2018).

Table 29: Titles of the selected papers from the 2nd Green and Sustainable Chemistry conference published in *Current Opinion in Green and Sustainable Chemistry*

• Plants as resources for organic molecules: Facing the green and sustainable future today
• Sustainable challenges on the moon
• Sustainable chemistry: A solution to the textile industry in a developing world
• Biomass-derived electrodes for flexible supercapacitors
• Recent Trends in Green and Sustainable Chemistry & Waste Valorisation: Rethinking Plastics in a circular economy
• Sustainable chemistry challenges from a developing country perspective: Education, plastic pollution, and beyond

The challenges of circular economy, social problems with the education on plastics and pollution, a reflection on the sustainable material flow necessary for the establishment of a manned Moon base, and much discussion about plants and biomass; virtually none of these touch upon the classic topics of green chemistry found in the green chemistry conferences in the 1990s and early 2000s. The difference is not merely due to the progress of knowledge but in the way of framing the problems of interest.

The idea that there was a frontier between green and sustainable chemistry and that the two did not necessarily cover exactly the same topic was embraced not just by Klaus Kümmerer, but also by the father of green chemistry, James Clark. Both chemists wrote together in 2016 a chapter "Green and Sustainable Chemistry" for a manual in sustainability science. The chapter explains that "Sustainable chemistry includes economical, social and other aspects related to manufacturing and application of chemicals and products. It aims not only at green synthesis or manufacturing of chemical products but also includes the contribution of such products to sustainability itself."[77] If Clark, one of the founders of the entire green chemistry movement, agrees with such differentiation between the two concepts, it clearly indicates that sustainable chemistry was becoming a field on its own.

More and more articles trying to highlight the difference between the two fields were published in the second half of the 2010s. In 2017, a group of French scholars published in *Green Chemistry* an article entitled "Sustainable chemistry: how to produce better and more from less?" It explained that: "If green

---

[77] Klaus Kümmerer and James Clark, "Green and Sustainable Chemistry" in *Sustainability Science. An Introduction*, ed. Harald Heinrichs, Pim Martens, Gerd Michelsen, and Arnim Wiek (Dordrecht: Springer, 2016): 43–60, 48.

chemistry and sustainable chemistry are often considered as synonyms, it is clear that differences of meaning do exist."[78] For the French researchers:

> Sustainable chemistry can be defined as the development of an even safer and more environmentally-friendly chemistry but one which also equally integrates the priorities of economic competitiveness and societal concerns. Sustainable chemistry is a complex equation which must ensure the longevity of the human, animal, and vegetable species whilst taking into consideration issues related to accessing different resources (carbon, water, metals), problems of access to energy, global warming, the exponential increase in the human population, for which chemistry must allow a serene development, the social and environmental impact of the value chain, and the erosion of biodiversity, while of course maintaining economic competitiveness to create profit and business.

This is a fantastically comprehensive definition succinctly summarising many of the concerns that have been expressed in a less concise way elsewhere. The insistence on global warming and biodiversity is to be particularly noted. The same year, also in 2017, one German scholar summarised the way sustainable chemistry has been described in the literature so far in the following way:

> – Sustainable chemistry contributes to a positive, long-term development in society, environment and economy. With new approaches and technologies it develops value-creating products and services for the needs of civil society.
> – Sustainable chemistry increasingly uses substances, materials and processes with the least possible adverse effects. Moreover, substitutes, alternative processes and recycling concepts are used, and natural resources are conserved. Thus, damage and impairments to human beings, ecosystems and resources are avoided.
> – Sustainable chemistry is based on a holistic approach, setting measurable targets for a continuous process of change. Scientific research and education for sustainable development in schools and vocational training serve as an important basis for this development.[79]

Friege's conceptualisation links sustainable chemistry to a very broad range of issues and considers it to be a 'holistic' approach, mirroring the vocabulary of the German sanfte Chemie from the late 1980s and early 1990s.

---

78 P. Marion, B. Bernela, A. Piccirilli et al., "Sustainable chemistry: how to produce better and more from less?," *Green Chemistry* 19 (2017): 4973–4989.
79 Henning Friege, "Sustainable Chemistry – A concept with important links to waste management," *Sustainable Chemistry and Pharmacy* 6 (2017): 57–60.

2.7. Formalising sustainable chemistry as a discipline on its own (2015–2021)

In the 2010s, sustainable chemistry emerged as an attempt to address the insufficiencies of green chemistry as a broader and all-embracing alternative. And yet, fleshing out a concept is not enough to attract international attention and obtain new acolytes. What is needed are institutions, a manifesto (similar to Anastas's and Warner's 1998 book) and, in particular, some sort of a catchy code of conduct equivalent to the 12 principles.

As for the institution, it has been created in Germany in 2015 as a fruit of collaboration between the Umweltbundesamt (the German federal environmental agency, roughly an equivalent of the EPA in the US) and the German Ministry of Environment. The new International Sustainable Chemistry Collaborative Centre (ISC3) became a leading German institution promoting sustainable solutions in the chemical industry. In terms of its ambitions, it can be compared to the American Green Chemistry Institute created in 1997. The ISC3 is responsible for organising conferences and seminars, it also funds research, and above all, it provides a platform for scholars interested in the challenges of sustainability in chemistry.[80] Certainly, a great many institutions, centres, laboratories, and other entities all over the world practise sustainable chemistry as well. So were many that practised green chemistry other than the Green Chemistry Institute. However, the Green Chemistry Institute was not like the others. Led by Anastas and his collaborators, it argued for a specific understanding of what green chemistry should be. The ISC3 is similar in this respect. Led by Klaus Kümmerer and his collaborators, it does not simply work on topics that count as sustainable chemistry, but its members actively work on improving and promoting sustainable chemistry as a separate theoretical framework.

As for some form of a manifesto of sustainable chemistry, there were a few attempts, often coming from researchers connected (more or less formally) to the ISC3 or published in the pages of the journals founded by Kümmerer. One of the most comprehensive overviews of the discipline's aspirations is presented in a German-speaking article published in *Sustainable Chemistry and Pharmacy* in 2019. Entitled "Das Konzept der Nachhaltigen Chemie: Schlüsselfaktoren für den Übergang zu einer nachhaltigen Entwicklung" [The Concept of Sustainable

---

80 The official website of the ISC3, history section, accessed 22/02/2022, https://www.isc3.org/en/about-isc3/history.html.

Chemistry: Key Factors in the Transition to Sustainable Development], it offers the following definition (in German in original) of sustainable chemistry:

> Sustainable chemistry contributes to the positive, long-term development of the society, the environment and the economy. With new approaches and technologies, it creates attractive products and services to answer the needs of the civil society. Sustainable chemistry increasingly uses substances, materials and processes that have possibly the smallest harmful effects, it uses substitutes, alternative processes and recycling strategies, and conserves natural resources. It prevents damage and impairment to people, ecosystems, and resources. Sustainable chemistry is based on a holistic approach and sets measurable goals for a process of continuous change. Scientific research and education for sustainable development in schools and vocational training are an important basis for it.[81]

The article makes use of the already existing literature on sustainability to frame its ideas. This is one of the important differences between the sustainable and green chemistry papers. The authors who advocate sustainable chemistry usually focus on the broad trajectory of sustainability and consider OECD and UN declarations, conferences, and summits as the formative moments of their project. Green chemistry partisans are usually much more EPA-centric in their descriptions of the history of the field and more limited to American sources.

However, the authors of the 2019 paper did not ignore the contributions of green chemistry. On the contrary, they incorporated the notion of green chemistry in their definitions and underlined its importance for their project. At the same time, they reiterated that sustainable chemistry was broader than green chemistry and that the two should not be conflated. Green chemistry was only one of many of what they called "key elements for sustainable chemistry" (it is important to note that, in a typical German manner, the article often refers to green chemistry in English instead of using the German equivalent "grüne Chemie").

The problems arise when the article defines seven guiding principles for sustainable chemistry (Table 30).

---

81 Christopher Blum, Dirk Bunke, Maximilian Hungsberg et al. "Das Konzept der Nachhaltigen Chemie: Schlüsselfaktoren für den Übergang zu einer nachhaltigen Entwicklung," *Sustainable Chemistry and Pharmacy* 13 (2019): 100140.

Table 30: Seven guiding principles for sustainable chemistry (Blum et al. 2019)

English translation	Original in German
1. Design and use of inherently safe chemicals 2. Development and use of alternative substances and processes 3. Reduction of the input [of problematic substances into the environment] 4. Conservation of natural resources 5. Promotion of recycling 6. Improvement of market opportunities 7. Raising awareness of social responsibility in the Companies	1. Die Gestaltung und Einsatz inhärent sicherer Chemikalien. 2. Die Entwicklung und Einsatz alternativer Stoffe und Verfahren. 3. Die Verringerung der Einträge. 4. Die Schonung natürlicher Ressourcen. 5. Die Förderung von Wiederverwertung 6. Die Verbesserung der Marktchancen. 7. Die Wahrnehmung der sozialen Verantwortung in den Unternehmen.

This list, while programmatic and outlining the core elements of the concept in a concise manner, is at least somewhat ambiguous if we compare it to the list of the 12 principles of green chemistry. In fact, sustainable chemistry principles 1 and 2, and to an extent 3 and 4, can be easily interpreted from Anastas and Warner's principles. The accents are slightly elsewhere and the message is clearer, but there is little added value. While sustainable chemistry principles 6 and 7 were not directly formulated in the 1998 book, Anastas, Warner, and their collaborators emphasised many times (some could argue that too many) the importance of economic viability for the success of green chemistry, as well as the fact that green chemistry has to be promoted and adopted on a voluntary basis by the industry. These were the unwritten principles of green chemistry, so their formalisation may be helpful but they do not differentiate green chemistry from sustainable chemistry in any meaningful manner. The only fully new principle concerns recycling, and yet even this principle was implied by many scholars interpreting green chemistry and it is hardly original. These seven principles bring little novelty to the practitioners of chemistry as compared with the green chemistry ones. This does not mean that the entire concept as presented in the article is not different from that of green chemistry (it is), or that it does not deserve attention, but that when it comes to concisely formalising it, the authors did not go far enough.

Perhaps a clearer formalisation of what makes sustainable chemistry particular was published in English in January 2021 by the German ISC3 in a 'dialogue paper' describing the "Key Characteristics of Sustainable Chemistry" (Table 31).[82]

Table 31: ISC3's key characteristics of sustainable chemistry (2021)

Holistic	Guiding the chemical science and the chemical sector towards contributing to Sustainability in agreement with sustainability principles and general understanding and appreciating potential interdependencies including long-distance interactions and temporal gaps between the chemical and other sectors.
Precautionary	Avoiding transfer of problems and costs into other domains, spheres and regions at the outset, preventing future legacies and taking care of the legacies of the past including linked responsibilities.
Systems thinking	Securing its interdisciplinary, multidisciplinary and transdisciplinary character including a strong disciplinary basis but taking into account other fields to meet Sustainability to its full extent. Application as for industrial practice including strategic and business planning, education, risk assessment and others including the social and economic spheres by all stakeholders.
Ethical and Social Responsibility	Adhering to value to all inhabitants of plant earth, the human rights, and welfare of all live, justice, the interest of vulnerable groups and promoting fair, inclusive, critical, and emancipatory approaches in all its fields including education, science, and technology.
Collaboration and Transparency	Fostering exchange, collaboration, and right to know of all stakeholders for improving the sustainability of business models, services, processes and products and linked decisions including ecological, social, and economic development on all levels. Avoiding all "green washing" and "sustainability washing" by full transparency in all scientific and business activities towards all stakeholders, and civil society.
Sustainable and Responsible Innovation	Transforming fully the chemical and allied industries from the molecular to the macroscopic levels of products, processes, functions and services in a proactive perspective towards sustainability including continuous trustworthy, transparent and traceable monitoring.

---

82 Klaus Kümmerer, Ann-Kathrin Amsel Amsel, and Dorota Bartkowiak, "Key Characteristics of Sustainable Chemistry. Towards a Common Understanding of Sustainable Chemistry," published in on 13/01/2021 on the ISC3's website, accessed 23/02/2022, https://www.isc3.org/fileadmin/user_upload/Documentations_Report_PDFs/ISC3_Sustainable_Chemistry_key_characteristics_20210113.pdf.

Sound Chemicals Management	Supporting the sound management of chemicals and waste throughout their whole life cycle avoiding toxicity, persistency and bio-accumulation and other harm of chemical substances, materials, processes, products and services to humans and the environment.
Circularity	Accounting for the opportunities and limitations of a circular economy including reducing total substance flows, material flows, product flows, and connected energy flows at all spatial and temporal scales and dimensions especially with respect to volume and complexity.
Green Chemistry	Meeting under sustainable chemistry application as many as possible of the 12 principles of green chemistry with hazard reduction at its core when chemicals are needed to deliver a service or function whenever and wherever this complies with sustainability
Life Cycle	Application of the above-mentioned key characteristics for the whole lifecycle of products, processes, functions and services on all levels, e.g. from molecular to the macroscopic levels and all sectors in a pro-active perspective towards sustainability

Table 31 constitutes the most comprehensive and self-explanatory overview of how sustainable chemistry practitioners see the core elements of their discipline, and the lines along which it is poised to evolve in the future. There are of course some interesting observations concerning these key characteristics and in order to fully appreciate them, a more rigorous exegesis would be necessary. The codification notably understands green chemistry in a rather narrow way, as the 12 principles, somewhat neglecting the fact they can be interpreted in an extremely generous manner. It does not mention the 12 principles of green engineering, nor the fact that for many green chemistry practitioners the LCA is an essential part of their activity. This is to say that the frontiers of these various concepts vary depending on whom you ask. At the same time, among these characteristics, many are truly unique to the sustainable chemistry project, such as for example the "collaboration and transparency" to avoid greenwashing, the "holistic" character of the overall ambition, and the "sound chemicals management" which implies close interdisciplinary collaborations. In general, of all existing conceptualisations, the ISC3's is the most capable of competing with the 12 principles of green chemistry as a vehicle to spread the idea of sustainable chemistry as something genuinely different and more embracing. Whether it stands the test of time and achieves similar popularity remains to be seen.

Ironically, however, in some respects, there are also less commendable tendencies in sustainable chemistry that mirror the development of green chemistry. What I mean here are the *post facto* generated genealogies trying to give the concept more historical legitimacy. One such attempt was published in 2021 in *Green Chemistry* (one of the co-authors was Kümmerer).[83] The article in question, "Education in green chemistry and in sustainable chemistry: perspectives towards sustainability," presents in a few paragraphs a short history of both green and sustainable chemistries. As for green chemistry, the paper reproduces the standard narrative I discussed in chapter 1: green chemistry as the product of the EPA and the 12 principles as its ultimate credo. However, when it comes to sustainable chemistry, the article places its birth in the early 1990s in the German Chemical Society. The article says that: "This approach was motivated by a broader view from the environment, i.e., end of products' life side, not from the synthesis side mainly."[84] This statement is at least slightly problematic. The life-cycle of products is a broad theme discussed by a variety of stakeholders in the 1990s all over the world. It has been mentioned in the early green chemistry papers, although it was not at the heart of the EPA's project (even if, as we remember, the rapprochement with industrial ecology suggests things were more complex). What matters is that nothing warrants the idea that Germans discussed it more thoroughly than Americans or anyone else. Moreover, nothing remotely similar to the formalisation of green chemistry principles took place in Germany in the context of sustainable chemistry. As already explained, there was Christ's Production-Integrated Environmental Protection, but it was very much focused on synthesis and similar to green chemistry's objectives, and there was sanfte Chemie, which was certainly very ambitious but outside mainstream science. Nachhaltige Chemie, if ever used, was a very intuitive concept. Comparing it with the EPA's green chemistry over the period is misleading and historically problematic.

In the following paragraphs of the 2021 paper, the authors invoke the OECD definition of sustainable chemistry from the 1999 publication as another founding moment for the new discipline. Again, not strictly speaking false, it has to be reiterated that this definition was not in dissent with the EPA's green chemistry and the majority of the participants at the OECD seminar did not differentiate the two at all, mostly because the EPA's way of thinking about green chemistry was still under the development, and the 12 principles have not yet conquered the imaginary of chemists all over the world.

---

83 Vânia G. Zuin, Ingo Eilks, Myriam Elschami, and Klaus Kümmerer, "Education in green chemistry and in sustainable chemistry: perspectives towards sustainability," *Green Chemistry* 23 (2021): 1594–1608.
84 Zuin, Eilks, Elschami, and Kümmerer, "Education in green chemistry," 1596.

Just as the early history of green chemistry has been smoothed down in dozens of publications by conveniently forgetting about the conceptual struggles of the 1990s and placing the founding date somewhat arbitrarily in 1991, the reconstruction of the history of sustainable chemistry in the 2021 paper plays a similar role. It attempts to recast sustainable chemistry as a tradition as old and venerable as green chemistry in order to build its own separate identity and justify its autonomy. The problem is that this reconstruction rests upon very shaky ground. It should be reiterated that while sustainable chemistry emerged organically over the last three decades, it was not until very recently (the early 2010s) that it acquired anything resembling a theoretical framework, very unlike the EPA's green chemistry that can be clearly traced to the 1990s, and whose own theoretical foundation was fully laid out in 1998.

At the margin of the discussion above, it is worth noting that this article's purpose was overall to show the superiority of the new idea:

> In summary, although GC [Green Chemistry] is an important building block for SC [Sustainable Chemistry], it is not necessarily sustainable, as GC does neither address possible implications of using renewable resources such as total substances, material, product and energy flows or alternative business models or catalysis trade-offs if metallic catalysts, are used, nor ethics or stakeholder roles. Furthermore, it is not clear how many of the 12 GC principles have to be met and how "ambitious" it has to be to call a chemical "green" or "greener" compared to another one.

This is a pretty harsh criticism to have been published in a journal whose name is *Green Chemistry*! But it perhaps shows that, at least among some scholars in the field, the problems of greenness are more and more obvious and the idea of new sustainable chemistry offers a welcome shift in focus.

Conclusions

What is sustainable chemistry? Similarly to green chemistry, its epistemological status is unclear. It covers activities that go beyond what counts as the science of chemistry, strictly speaking, with its focus on social dialogue and 'non-chemical aspects of chemistry.' It is certainly holistic in its ambition. This ambiguous nature was the reason it took so long to formalise its foundations. If green chemistry had its clear founding moments, then sustainable chemistry was more of a background concept slowly 'hatching' for more than 20 years before being embraced more explicitly by chemists as an alternative to green chemistry.

If 'hatching' is the metaphor to use, it is worth talking about the conditions that made it possible. In fact, these were the particularities of the German-speaking world that propelled the development of this framework. The rejection of the green terminology due to political connotations and the embrace of the sustainability language created a unique environment, in which new modes of thinking could flourish. What is interesting about it, is that such development could not have been planned. On the contrary, German chemical industry's reluctance concerning the term "grüne Chemie" could be even seen as an obstacle to transforming it along more environmental lines. And yet the opposite happened and this is the German sustainable chemistry that is larger and more radical in scope than what came to be traditionally understood as green chemistry.

Bringing under one umbrella green chemical synthesis, life cycle assessment, and chemical leasing is a bold step. However, it remains unclear how the framework will develop in the coming years. If it is to be meaningful, it needs institutions that would at least attempt to integrate these different dimensions in their work, making chemists collaborate in an interdisciplinary environment. The risk is that, with limited funding and external constraints, it will be just easier to fall back on the more familiar realm of chemical synthesis. In other words, sustainability chemists would preach to the world ideas they do not really practise. Sustainable chemistry requires rethinking what we mean by laboratories by creating more integrated and open workspaces welcoming the input of various stakeholders. Again, industrial ecology may be the key discipline to build on, and the experience from the older frameworks such as sanfte Chemie may prove itself crucial for sustainable chemistry to succeed. To sum up, even if sustainable chemistry may be a strong contender to overthrow green chemistry as the synonym of environmental thinking in chemistry, this evolution is, as of now, far from certain.

Chapter 8: New conceptual frontiers for chemistry and environment

The EPA's green chemistry came to single-handedly dominate the discourse of environmentally-minded chemists in the early twenty-first century. If you practise chemistry or make funding decisions about chemical research, but you want (or have to) incorporate some environmental accountability, you will almost certainly need to relate to the green chemistry concept or to the 12 principles. Sustainable chemistry is nevertheless emerging as a valid alternative. Will it come to supersede the green one? Its all-embracing character may be seen as an advantage.

And yet in the jungle of scientific concepts, the ruthless rule of the survival of the fittest makes the law. From the melting pot of ideas emerge continuously new concepts which result either from the unfamiliarity or from the discontentment with green and sustainable chemistries. They usually position themselves as prolongations of some previous traditions or surf on the popularity of fashionable buzzwords. Every single one of them can become the new green chemistry of the future if it catches the wind in its sails and gets some official institutional backing.

On top of that, green chemistry has not yet said the final word. As of 2020, the number of articles tagged as green chemistry was ten times higher than as sustainable chemistry. While the theorists of sustainable chemistry gladly incorporate green chemistry into their framework by underlining that the latter is way too narrow to address the environmental challenges of the future, this is also because sustainable chemistry partisans interpret green chemistry in a narrow way. And yet the concept substantially changed its meaning throughout its lifetime.

In this chapter, I discuss new alternative concepts fighting for recognition against green and sustainable chemistry, along with some of the most recent shifts in understanding what hides behind the green chemistry concept. It is perhaps impossible to map all these different conceptualisations and give them full justice (at least without simply reproducing the foundational articles in their integrality). The purpose of this chapter is then to identify a few of them in order to give a sense of what is happening at the frontiers of the mentalities of modern green chemists.

1. New contenders to overthrow green chemistry

1.1. Conservative evolution and the risk of politicization of green chemistry debates

Every now and then, researchers formulate novel ways of thinking by giving a new spin to the theoretical concepts that had been already well established. Sometimes the results can be groundbreaking (that was the case of the 12 principles of green chemistry which superposed many disparate lines of thinking in a single framework), but sometimes the foundations on which the new concept is built are shaky or refer to ideas many chemists would not immediately subscribe to. The problem is that with plethora of new frameworks being forged every year, some that slip through peer-review happen to be highly unconventional. I chose to present one such paper, neither because it was particularly influential nor because it is poised to play any role in the future, but precisely because of its original theoretical framing. What many good-willed chemists forget is that when they try to extend the frontiers of their discipline (be it green or sustainable chemistry) and make it more 'socially relevant,' or when they argue for new forms of chemical ethics, they enter the murky ground of politics and ideology.

In 2018, the *Green Chemistry* journal published a paper entitled "Conservative Evolution and Industrial Metabolism in Green Chemistry" contextualizing much of recent green chemistry research within the two frameworks from the title.[1] The Hungarian authors of the paper define them in the following way:

> (i) conservative evolution ... refers to the observation that throughout the history of the universe, old constructs such as elementary particles, amino acids, or living cells remained conserved while the world evolved in its complexity and (ii) the practice of industrial metabolism ... is a manifestation of conservative evolution in human activity referring to the application of biological metabolism in the production of goods for our civilization.

None of these two concepts originated in the chemical community. Whereas industrial metabolism is a genuine well-established idea deeply rooted in the literature on industrial ecology, the concept of conservative evolution is considerably more esoteric and not grounded in the existing scientific literature. Therefore defining industrial metabolism as a manifestation of conservative evolution is a rather unconventional choice. To make things even more bizarre, the authors

---

[1] Gábor Náray-Szabóa and László T. Mika, "Conservative Evolution and Industrial Metabolism in Green Chemistry," *Green Chemistry* 20 (2018): 2171–2191.

define evolution not in Darwinian terms but "as a process, which leads to an increased and viable complexity." The footnote to this statement refers to the works of the extravagant Catholic theologian Teilhard de Chardin, one of the godfathers of the Second Vatican Council, famous for claiming that the universe evolves towards the Omega point he identified with Jesus Christ (the Logos).

To explain the "conservative evolution" further, the authors refer to a single 2014 publication under the title "Conservative Evolution, Sustainability, and Culture" authored by one of them in a small humanities journal *Comparative Literature and Culture*.[2] This publication is a rather eclectic essay superposing themes as different as the development of matter after the Big Bang, the biological evolution, and the transformations of human cultures. The author refers to Lovelock's controversial Gaia hypothesis (positing that Earth is a self-regulating homeostatic system enabling life through unconscious feedback), to the already mentioned Chardin, and to many other original thinkers. The argument is that there is an underlying principle of conservative evolution that works similarly on different levels of organization:

> Evolution occurs by the adaptation of existing mechanisms to new purposes. Thus, the air bladder of a fish becomes the lung of an amphibian and a feudal parliament that evolved to resolve disputes between nobles and king becomes an instrument of democracy. It seems to me that evolution is a process wherein continuous replacement of the old is based on well-established structures.[3]

The entire connection between conservative evolution, industrial metabolism, and green chemistry is not very well articulated in the 2018 paper and the first term appears only in the introduction and the conclusions. However, the sole mention of the concept indicates a very serious challenge. The more green and sustainable chemistry embraces social considerations, the more the ideological and political discourses penetrate the publications in the field. Sustainability itself has different definitions and partisans. Some see it as an extension of the Catholic social teaching,[4] others see it as one of the tools for constructing a new

---

2 Gábor Náray-Szabó, "Conservative Evolution, Sustainability, and Culture," *Comparative Literature and Culture*, 16 (2014): http://dx.doi.org/10.7771/1481-4374.2316.
3 Náray-Szabó, "Conservative Evolution, Sustainability, and Culture," 5.
4 Ian Christie, Richard M. Gunton and Adam P. Hejnowicz, "Sustainability and the common good: Catholic Social Teaching and 'Integral Ecology' as contributions to a framework of social values for sustainability transitions," *Sustainability Science* 14 (2019): 1343–1354.

socialist society.[5] And who is to tell that Teilhard de Chardin is not an early herald of sustainability? Are the reviewers of *Green Chemistry* equipped to confront these questions? While it would be easy to dismiss the question as marginal, it is essential to point out that politics is in the genetic code of green and sustainable chemistry, and sooner or later this type of problems may resurface. After all, the conservative evolution paper was published in the leading journal in the field. Would the editorial committee accept a paper recasting green chemistry as, for example, the "people's chemistry" through a Marxist lens? It is worth reminding that the sanfte Chemie project was strongly rooted in the writings of the Frankfurt School of critical philosophy. Another, perhaps somewhat less controversial but no less problematic, example comes from a 2016 editorial in *Green Chemistry* published under the title "Reflection and perspective on green chemistry development for chemical synthesis—Daoist insights," which tried to link principles of green chemistry to an ancient Chinese religious philosophy of Lao Tzu.[6] Again, as anecdotal these rare publications may seem, the underlying problem of appropriation of greenness by various ideological and religious currents should not be dismissed.

1.2. One-world chemistry

Of course, not all novel concepts are that politically charged. On the contrary, some present non-controversial aspirations in the spirit of the good old universalism. In 2016, a high-profile journal *Nature Chemistry* published an article entitled "One-world chemistry and systems thinking." The authors set themselves an ambitious goal: "The practice and overarching mission of chemistry need a major overhaul in order to be fit for purpose in the twenty-first century and beyond."[7] Reformulating this statement in Kuhnian terms, there was a need for a paradigm shift in chemistry! Why?

> (1) the discipline has not been effective in reinventing itself or projecting its contemporary advances on prominent external platforms, (2) it is intrinsically an incremental science and has acquired the image of being a jaded and slow-paced discipline in an era

---

5 David S. Pena, "The Six Essential Components of Sustainable Socialism: From Building the Productive Forces to Combating Bourgeois Liberalization," *International Critical Thought* 4 (2014): 267–288.
6 Chao-Jun Li, "Reflection and perspective on green chemistry development for chemical synthesis—Daoist insights," *Green Chemistry* 18 (2016): 1836–1838.
7 Stephen Matlin, Goverdhan Mehta, Henning Hopf, and Alain Krief, "One-world chemistry and systems thinking," *Nature Chemistry* 8 (2016): 393–398.

of 'big bang' or 'instant' science, (3) it has witnessed encroachment on its own core space by other disciplines (for example, nanoscience, molecular biology and Earth sciences) and (4) there has been a lack of communication and public engagement, lack of sensitivity to societal issues, and an absence of credible and influential voices to articulate chemistry's importance and contributions.

These are certainly hefty accusations. They resemble slogans from the 1990s publications in which the bad public image of chemistry was one of the reasons motivating the reinvention of the discipline with green tools. Some of them may sound controversial though: is chemistry not being an 'instant science' a bad thing? Perhaps it provides a slower but more systematic approach to solving global problems? Does chemistry really suffer from 'encroachments' from other disciplines? Or perhaps interdisciplinary cross-fertilization is just how science works today? There are some important questions about the underlying assumptions lying behind the concept, but I rather focus on the paper's core message: the grand challenges the one-world chemistry is supposed to solve.

> In our view, the re-imagined chemistry must go beyond 'being a science' and embrace the concept of 'being a science for the benefit of society.' This implies that chemistry must pursue a triple role, involving (1) creating new scientific knowledge, (2) translating knowledge into useful applications and (3) helping to meet the emergent challenges of multiple unfolding global crises. Chemists have traditionally engaged more in the first two elements, but adding a strong emphasis to the third is now an imperative.

This reimagined chemistry is called the "one-world chemistry" that

> embodies the idea that chemistry is a creative science that is practiced in both fundamental and applied arenas in a sustainable and ethical manner for the benefit of society. A central characteristic must be that it enables those who learn and practice chemistry to be aware of and respond to the interconnectedness of chemistry and related chemical sciences with local and global systems.

Two key elements are reiterated in the text: the one-world chemistry is meant as a new sustainability science (a claim that was made by industrial ecology scholars already back in the early 1990s), as well as a trans-disciplinary systems science. The authors provide a table to summarize key points of what they see as one-world chemistry (Table 32).

Table 32: The concept of one-world chemistry[8]

Roles	Goals	Approaches	Orientations
-Creating new scientific knowledge -Translating knowledge into useful applications -Helping to meet the challenges of multiple unfolding crises	-Being a science for the benefit of society-oriented -Being a 'sustainability science' -Being an ethical science	-Adopting sustainability principles -Embracing systems thinking -Working across disciplines -Strengthening the productive interface with industry -Taking an ethical approach	-Internal and External: Aligning the internal content, practice and teaching of chemistry and its external orientation, connections and engagements with other disciplines and the world at large with the goals and approaches of one-world chemistry -External: Engaging strongly with the public and media, projecting chemistry's creativity, its past contributions and its future potential to be a science for the benefit of society

What is interesting is the foundation from which this new paradigm is supposed to develop. If green chemistry was a result of the pollution prevention philosophy, sustainable chemistry was the translation of sustainability into the realm of chemistry, the one-world chemistry scholars ground their concept in the "Hague Ethical Guidelines for the practice of chemistry" developed in 2015 and endorsed by IUPAC in 2016.[9] In addition to sustainability and systems thinking, "ethics" and "social development" seem to be the keywords for understanding the aspirations of the project.

Who are the proponents of one-world chemistry? The concept was coined inside the UNESCO-endorsed International Organization for Chemical Sciences in Development (IOCD) created in 1980 which defines its mission with two points: "Repositioning chemistry as a science for the benefit of society" and "Promoting the role of the chemical sciences in sustainable development."[10] The

---

8 Matlin, Mehta, Hopf and Krief, "One-world chemistry and systems thinking," 394.
9 Hague Ethical Guidelines, accessed 23/02/2022, https://www.opcw.org/hague-ethical-guidelines.
10 IOCD's website, mission section, accessed 23/02/2022, http://www.iocd.org/About/visionMission.shtml.

IOCD advertises the one-world chemistry as one of the institution's defining ambitions on its website.[11] As such, this novel concept is both the fruit of the IOCD's anterior reflection and the program for its future actions.

In practice, however, the concept of one-world chemistry did not much develop since 2016. The problem is that if one-world chemistry describes itself as the science of sustainability, it puts itself in a somewhat awkward position towards sustainable chemistry and the situation begs the question whether there is a difference between the two, the genealogy aside. Stephen A. Matlin, the father of one-world chemistry wrote in 2020 that: "Sustainable chemistry is a broader field that includes 'green chemistry'. It adopts systems thinking and cross disciplinary approaches, as advocated in 'one-world' chemistry, and new business models to encompass the entire life cycle of materials and reduce the use of all resources and materials flows."[12] This passage does not explain, however, what the real difference between the two frameworks would be. Certainly, Matlin insists on the importance of systems science more than sustainable chemistry scholars. But was not systems thinking already a part of sustainable chemistry for quite some time, even if not always in an explicit way? Not to even mention that the insistence on 'systems thinking' fails to account for the fact that the systems theory was constitutive for the discipline of environmental chemistry already back in the 1970s; a topic I discuss more in detail in the second part of this chapter. The point is that many of the aspirations of one-world chemistry have been expressed more skilfully many times before. It appears as if the one-world chemistry founding fathers failed to realise in 2016, when they published their original paper, that both green and sustainable chemistries had been around for quite a while. It remains unclear what would its added value be in comparison with these older approaches and none of the later papers concerning one-world chemistry actually addresses the issue by connecting in any manner to the vast scholarship on either green or sustainable chemistry.

And yet one-world chemistry is not the only such supposedly revolutionary concept trying to reframe chemical practice that has been published in recent years. In fact, Matlin mentions in 2020 one more framework struggling for recognition: circular chemistry.

---

11 IOCD's website, introductory section, accessed 23/02/2022, http://www.iocd.org/OWC/intro.shtml.
12 Stephen Matlin, Sustainability and chemistry, 2020, published on the IOCD's website, accessed 23/02/2022, http://www.iocd.org/perspectives/606-2020-IOCDperspectve-SustainabilAndChem-13p.pdf.

1.3. Circular chemistry

Circular chemistry has a distinct pedigree as it takes after its mother concept: the circular economy. Narratives of its history usually agree that the concept of circular economy had some antecedents throughout the twentieth century and that the term itself was used for the first time in the 1990s, but that it entered into the language of public stakeholders only in the early 2010s thanks to the report *Towards the Circular Economy: Economic and Business Rationale for an Accelerated Transition* commissioned by the Ellen MacArthur Foundation.[13] The concept became a part of the EU's economic strategy in the following years. According to European Parliament's website:

> The circular economy is a model of production and consumption, which involves sharing, leasing, reusing, repairing, refurbishing and recycling existing materials and products as long as possible. In this way, the life cycle of products is extended.
> In practice, it implies reducing waste to a minimum. When a product reaches the end of its life, its materials are kept within the economy wherever possible. These can be productively used again and again, thereby creating further value.
> This is a departure from the traditional, linear economic model, which is based on a take-make-consume-throw away pattern. This model relies on large quantities of cheap, easily accessible materials and energy.[14]

I refrain from engaging in the discussion on the advantages and limitations of this new framework and its relation with the more established ones such as industrial ecology. It may be argued that the core idea of industrial ecology is, in fact, to build a circular economy and there is little added value in the latter as a separate framework. Whatever our opinion may be, the circular economy became a major trend around 2015 (Figure 31).

---

[13] Ellen McArthur Foundation's website, circular economy section, accessed 23/02/2022, https://ellenmacarthurfoundation.org/towards-the-circular-economy-vol-1-an-economic-and-business-rationale-for-an.

[14] European Parliament's website, circular economy definition, accessed 23/02/2022, https://www.europarl.europa.eu/news/en/headlines/economy/20151201STO05603/circular-economy-definition-importance-and-benefits.

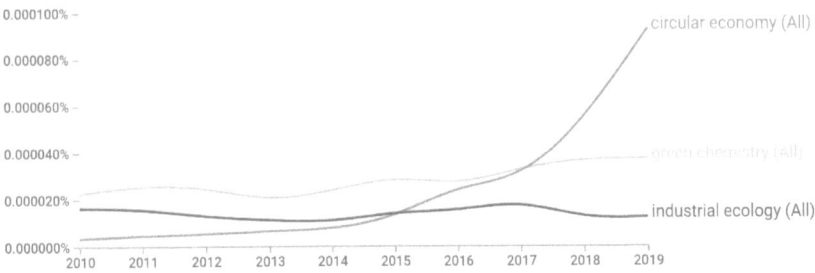

Figure 31: Popularity of terms "circular economy," "industrial ecology" and "green chemistry" according to google Ngram (2010–2019, smoothing = 0)

Green chemists took notice of this new fashionable idea. Articles on how green chemistry can contribute to a circular economy became plentiful in the second half of the 2010s. Their titles tell us a lot about their ambition: "Is Green Chemistry a feasible tool for the implementation of a circular economy?," "Implementation of green chemistry principles in circular economy system towards sustainable development goals: Challenges and perspectives," "Green chemistry and the plastic pollution challenge: towards a circular economy," "Ripe for disruption: reimagining the role of green chemistry in a circular economy."[15] All of them try to fit the already established concept of green chemistry into the rapidly growing framework of circular economy. And yet, of course, it does not fit exactly. Recycling is perhaps the least ingrained idea in green chemistry; that was not what green chemistry was about for many years. Instead of retrofitting recycling and circularity under the green chemistry umbrella, why not invent a new more appropriate framework?

---

15 Natalia Loste, Esther Roldán, and Beatriz Giner, "Is Green Chemistry a feasible tool for the implementation of a circular economy?," *Environmental Science and Pollution Research* 27 (2020): 6215–6227; Tse-Lun Chen, Hyunook Kim, Shu-Yuan Pan et al., "Implementation of green chemistry principles in circular economy system towards sustainable development goals: Challenges and perspectives," *Science of the Total Environment* (10 May 2020), 716: 136998, doi: 10.1016/j.scitotenv.2020.136998; Roger A. Sheldon and Michael Norton, "Green chemistry and the plastic pollution challenge: towards a circular economy," *Green Chemistry* 22 (2020): 6310–6322; Mats Linder, "Ripe for disruption: reimagining the role of green chemistry in a circular economy," *Green Chemistry Letters and Reviews* 10 (2017): 428–435.

That is what three Dutch chemists from the University of Amsterdam did in 2019 in an article published in *Nature Chemistry* under the title "Circular chemistry to enable a circular economy."[16] The article announces:

> By expanding the scope of sustainability to the entire lifecycle of chemical products, the concept of circular chemistry aims to replace today's linear 'take–make–dispose' approach with circular processes. This will optimize resource efficiency across chemical value chains and enable a closed-loop, waste-free chemical industry.

Do authors suggest here that sustainability did not apply to the entire lifecycle of chemical products before? As such, the relation of this paper to the already existing scholarship concerning the life-cycle assessment remains somewhat unclear. Furthermore, the authors seem to ignore the existence of sustainable chemistry as a separate tradition, just like the authors of the one-world chemistry paper (in 2016 at least).

Unlike the one-world chemistry paper, however, this one refers to green chemistry, although in a quite unusual manner: "Since it was first introduced in the 1980s, green chemistry has provided a framework for teaching and performing sustainable chemistry, and has delivered an impetus for developing cleaner products and processes." Leaving aside the peculiar choice for the date of birth (why the 1980s?) and the fact the authors conflate green and sustainable chemistry (do they mean that green chemistry methods helped to make chemistry more sustainable?), they recognize green chemistry's added value and their article reproduces the entire set of the 12 principles as formulated by Anastas and Warner. And yet, they continue, green chemistry is "perfectly suited for the optimization of linear production routes" but the "developments towards a circular economy … require a re-evaluation of what defines a sustainable chemical process." There is, according to the authors, a need for some sort of a new framework to accommodate this new economic reality.

To show that green chemistry is not enough to achieve this ambition, they offer two case studies to prove their point. These case studies are, however, rather peculiar. The first concerns the green synthesis of adipic acid.

> Solvent-free conditions are applied (GC 2) [Green Chemistry Principle 2], avoiding the use of the corrosive nitric acid (GC 3) and thus side-stepping the formation of the environmentally taxing gas nitrous oxide, N2O — a waste product of the current industrial synthesis (GC 4). The green method, however, requires hydrogen peroxide, H2O2, as

---

16 Tom Keijer, Vincent Bakker and J. Chris Slootweg, "Circular chemistry to enable a circular economy," *Nature Chemistry* 11 (2019): 190–195.

starting material, which means that this process is currently not commercially viable, since H2O2 is more expensive than the adipic acid product. Although this route obeys green chemistry principles, it violates the value chain.

The case study refers in the footnote to a 1998 article published in *Science*: "A "Green" Route to Adipic Acid: Direct Oxidation of Cyclohexenes with 30 Percent Hydrogen Peroxide."[17] This choice prompts many questions. Why refer to a 20-year-old article that while may have "green" in the title, does not connect to the EPA's green chemistry framework (or to any framework for that matter)? Why analyse it through the lenses of the 12 principles if the article itself does not make use of the 12 principles, and its authors were probably not even aware of their existence at that time? There has been a substantial amount of scholarship in green chemistry published over the last decades, and it would be certainly easy to cherry-pick some article pretending to be green to prove any point, and yet, despite that, the authors of the circular chemistry paper chose specifically an article that is absolutely not representative for green chemistry as a discipline. Yet perhaps we should not focus on the case study itself, but on the overall argument the authors are trying to make.

The message this case study was supposed to transmit is that green chemistry may be good for the environment, but is not necessarily economically viable. However, as already explained in chapters 1 and 2, from the very outset of the green chemistry project, the industry was pandered to and the importance of commercial viability was heralded as the necessity for any green synthesis to be adopted. Business friendliness was an unwritten 13th principle of green chemistry. If anything, green chemistry was overly accommodating towards the industry by focusing exclusively on research that would increase profits, instead of exploring all possible routes that might be less commercially interesting but much more sustainable and environmentally friendly. Does it mean that the authors of the 2019 paper believe that their circular chemistry should be *even more* profit-oriented than green chemistry?

The authors provide us with one more unusual case study. The second failure of the green chemistry principles, according to the fathers of circular chemistry, is illustrated with the Haber-Bosch process which is "economically viable, yet remain unsustainable."

---

17 K. Sato, M. Aoki, and R. Noyori, "A "Green" route to adipic acid: direct oxidation of cyclohexenes with 30 percent hydrogen peroxide," *Science* 281 (1998): 1646–1647.

It is a key industry showcase for the use of catalysts (GC 9) in increasing energy efficiency (GC 6). The current process requires high temperatures and pressures, and further optimization has stagnated. After its use as fertilizer, large portions of the fixated nitrogen are lost to the environment, causing eutrophication, a global environmental concern … The cascade of environmental changes that results includes an increase in water and air pollution, both of which threaten to destabilize the Earth's system beyond the proposed 'planetary boundaries' or 'safe operating space' for anthropogenic activities.

This argument is particularly incoherent. While retrofitting some chemical processes into the green chemistry framework may be used to prove that the brown/green divide is somewhat artificial, the entire point of the green chemistry approach was to reject 'the old ways,' disqualifying the Haber-Bosch as green chemistry from the start. Besides, even if we adopted the 12 principles as the ultimate reference for greenness ignoring the historical context, the Haber-Bosch could hardly be considered green since the authors cite only two out of the twelve principles, while, as already explained, Anastas and his collaborators repeatedly emphasised that they should be considered holistically, and that one cannot call green a process that satisfies some principles but completely fails the others. Finally, if we turned the argument on its head, for someone more sustainability-oriented, it may be argued that the Haber-Bosch process was much more sustainable than the previous solution involving the transportation of guano from around the world to Europe and using massive amounts of sulphuric acid to manufacture superphosphates for agriculture. If our objective is to make things more sustainable in the long run, the Haber-Bosch played perhaps a positive role in addressing the great challenge of feeding humanity. Overall, no matter which perspective we adopt and from whatever side we analyse the Haber-Bosch process, the case against it, as presented in the 2019 article, does not stand. It is simply completely irrelevant to the argument the authors are trying to make and betrays the lack of understanding of the green chemistry framework.

Whatever we make of these case studies, the paper wants to address these perceived limitations of green chemistry and argues for establishing the new circular chemistry approach which "promotes, in particular, resource efficiency across chemical value chains and highlights the need to develop novel chemical reactions to reuse and recycle chemicals, to in turn enable development towards a closed-loop, waste-free chemical industry." If systems thinking was a guiding term for one-world chemistry, for circular chemistry the equivalent is the term 'waste.' To achieve their ambitions, the Dutch chemists formulated the 12 principles of circular chemistry that are not meant to replace but to complement the 12 principles of green chemistry (table 33).

Table 33: 12 principles of circular chemistry (2019)

1. Collect and use waste.	Waste is a valuable resource that should be transformed into marketable products.
2. Maximize atom circulation	Circular processes should aim to maximize the utility of all atoms in existing molecules.
3. Optimize resource efficiency	Resource conservation should be targeted, promoting reuse and preserving finite feedstocks.
4. Strive for energy persistence	Energy efficiency should be maximized.
5. Enhance process efficiency	Innovations should continuously improve in- and post-process reuse and recycling, preferably on-site.
6. No out-of-plant toxicity	Chemical processes should not release any toxic compounds into the environment.
7. Target optimal design	Design should be based on the highest end-of-life options, accounting for separation, purification and degradation.
8. Assess sustainability	Environmental assessments (typified by the LCA) should become prevalent to identify inefficiencies in chemical processes.
9. Apply ladder of circularity.	The end of life options for a product should strive for the highest possibilities on the ladder of circularity.
10. Sell service, not product.	Producers should employ service-based business models such as chemical leasing, promoting efficiency over production rate.
11. Reject lock-in.	Business and regulatory environment should be flexible to allow the implementation of innovations.
12. Unify industry and provide coherent policy framework.	The industry and policy should be unified to create an optimal environment to enable circularity in chemical processes.

These principles offer some insights that were lacking in previous conceptualisations. The explicit reference to the LCA and the service-not-product approach are some valuable examples. At the same time, both the LCA and chemical leasing have been part of the sustainable chemistry strategies for years, not to mention the re-use of wastes and resource efficiency. Some of these principles borderline overlap with the principles of green chemistry and of green engineering (e.g. principles 2, 3, 4, or 7 of circular chemistry). Others, such as principles 11 and 12, are very vague and it remains unclear how they differ from the existing practices, aside from the fact they may constitute another injunction against environmental regulation and in favour of business friendliness. As such, these last two principles can be actually applicable to any chemical or scientific endeavour. The only genuine innovation on the list is the ladder of circularity (principle 9). Traditionally, the ladder was made of three steps: reduce, reuse, recycle. The famous three R's are the canon of sustainability in any manufacturing company. What the circular chemistry paper proposes are the 11 R's: reject, reduce, reuse, redistribute, repair, refurbish, repurpose, remanufacture, recycle, recover, return (to the environment).

The concept of circular chemistry gained some following in the literature in chemistry, at least more than one-world chemistry, and some scholars repurposed it for their own theses.[18] If specialists in circular economy embrace it, circular chemistry may successfully develop into something more important. The main problem nevertheless remains identical to the one already identified in the case of one-world chemistry. Is it really that much different from sustainable chemistry? Does not sustainable chemistry accommodate everything circular chemistry already advocates in favour of?

In a 2020 article published in *Science* by Kümmerer, Clark, and Zuin, under the title "Rethinking chemistry for a circular economy," these prominent chemists proposed their own framework connecting to the notion of circular economy.[19] While the term "sustainable chemistry" does not appear a single time in the text, the conceptual and institutional parentage of the reflection leaves no doubt. The article completely ignores the 2019 publication and its 12 principles of circular chemistry and attempts to build an alternative set of 15 guidelines for "Integrating chemistry into a circular economy" (Table 34).

---

18   S. Venkata Mohan and Ranaprathap Katakojwala, "The circular chemistry conceptual framework: A way forward to sustainability in industry 4.0," *Current Opinion in Green and Sustainable Chemistry* 28 (2021): 100434; Gregory Chatel, "Chemists around the World, Take Your Part in the Circular Economy!," *Chemistry. A European Journal* 26 (2020): 9665–9673.

19   Klaus Kümmerer, James H. Clark, and Vânia G. Zuin, "Rethinking chemistry for a circular economy," *Science* 367 (2020): 369–370.

Table 34: Guidelines for integrating chemistry into a circular economy

1. Keep molecular complexity to the minimum required for the desired performance, including end of life (complex molecules require more synthesis steps, may have additional undesirable properties, and can be more difficult to recycle).
2. Design products for recycling, including all additives and other components of the product.
3. Reduce and simplify diversity and dynamics of substance, material, and product flows; e.g., use fewer chemicals overall (both number and quantity), design for less resource intensity, and adapt innovation speed of products to adaptation speed of recycling.
4. Avoid complex products (e.g., multiple components, materials).
5. Minimize use of product components that cannot easily be separated and recycled (e.g., solvents, metals).
6. Design products not suitable for capture and recycling for complete fast mineralization at the end of their lives (e.g., pharmaceuticals, pesticides, personal care and cleaning products).
7. Prevent raw materials from becoming critical through reduced use and efficient recovery and recycling (e.g., many metals).
8. Avoid entropic losses and transfers(e.g., dissipation of metals, energy).
9. Avoid rebound effects (e.g., using less carbon often means higher demand for metals).
10. Be responsible for/develop ownership of your product throughout its complete life cycle, including recycling.
11. Ensure traceability and consider use of product digital passports (e.g., composition of products, components, and processes).
12. Develop and apply circular metrics (e.g., giving credit to the use of by-products).
13. Change traditional chemical practices based on "bigger-faster" into "optimal adapted-better-safer" and change ownership to rent, lease, and share business models.
14. Keep processes as simple as possible with a minimum number of steps, auxiliaries, energy, and unit operations (e.g., separations, purification).
15. Design processes for optimal material recovery of auxiliaries, unused substrates, and unintended by-products (based on quality and quantity).

These guidelines, while preceded by a less ambitious build-up than the circular chemistry principles, are their direct competitors. This list is what the most reputable sustainable and green chemistry representatives contribute to the debate on circular economy. Which list will be more cited in 10, 20, and 30 years? Which will have a more meaningful impact on the practices of chemists and of the chemical industry? Behind sustainable chemistry there is a community, institutions, and since recently, increasingly robust theoretical backing. Behind circular chemistry there is a catchy name (a factor of prime importance as proven by the 12 principles of green chemistry) and an exponentially expanding field of circular economy in general.

1.4. Concluding remarks on new alternative frameworks

If conservative evolution appears to be a rather personal project at the margin of the chemical community, both one-world chemistry and circular chemistry have not said their last words. They have all-embracing aspirations: to transform the way we think about chemistry. There are lingering questions however: do we really need them? Is there any added value to these approaches? This is not to criticize their ambitions and underlying ideas. Gathering and linking together chemical synthesis, systems thinking, and life-cycle assessment is an important initiative, but all of this has been already discussed back in the 1990s when industrial ecology dialogued with green chemistry. Some scholars proposing new concepts appear to ignore a mass of theoretical literature that has been published on these topics. Writing about environmentally friendly chemistry today as if nothing has changed between the early 1990s and 2020s can easily lead to reinventing the wheel.

There is a caveat though. After all, much of the criticism I addressed against green chemistry in the first chapters of this book comes from the ignorance of the field's own roots. The EPA's green chemistry was purposefully presented as a radical departure from previous approaches and its creators ignored the existence of similar ideas in the past committing the same sin as the fathers of one-world and circular chemistries. Moreover, green chemistry pioneers were unfamiliar with alternative approaches (especially of European pedigree) and brought together concepts that were completely unrelated to each other before (e.g. renewability, pollution prevention, laboratory safety). And yet green chemistry, due to its elegance, simplicity, but also vagueness, gained international fame, even if for the historian its foundational narrative is naive at best. As such, it is then impossible to know whether one of these new radical departures (be it one-world or circular chemistry, or something else altogether) will not restructure the entire discipline once again by capturing the imagination of chemists all over the world. The rise of new scientific stakeholders in China and India leaves this option possible.

At the same time, this may not happen at all, because green chemistry is in the middle of reinventing itself along the new lines as well.

2. New ideas for green chemistry

How to explain the fact that green chemistry, a somewhat simplistic idea of pollution prevention developed by the EPA in the 1990s, is today more popular than ever? The incessant promotion of the concept by Anastas, Warner, Clark, Sheldon, and many other scholars certainly contributed to its success. However, this success was also due to the malleability of the concept itself and to the openness of the founding fathers towards other fields of inquiry. Green chemists have always actively participated in the dialogue with the larger sustainability community; it has always been an ecumenical initiative. By growing, it incorporated more and more different concepts, trajectories, and frameworks. Principle 7 led green chemists to fully embrace the importance of renewability of raw materials, with a particular emphasis on biomass, even though this line of inquiry was barely mentioned in the first green chemistry publications. The life-cycle assessment, which was introduced to green chemists by industrial ecology scholars, became by the end of the 2010s the best criterion for evaluating the greenness of chemical reactions. In other words, green chemistry covers today many topics that were outside of its original scope.

In this short section, I want to point out two recent front line cases in which green chemistry engaged in one way or another with problems that were traditionally beyond its borders. These are individual instances, too early to be considered patterns foreshadowing things to come, but both of them highlight deep unresolved problems that green chemistry, in the broadest possible sense, is poised to encounter.

2.1. Systems thinking and green and sustainable chemistry

In 2017, IUPAC announced a project on integrating systems thinking into chemistry education.[20] It was a sign of a larger trend gaining more and more popularity over the last few years among environmentally-minded chemists; the one-world chemistry approach putting systems thinking at the heart of its philosophy is just one example. Systems theory has remained nevertheless largely absent in the pages of *Green Chemistry* and sister journals so far. Scholars identifying with sustainable chemistry played with the idea more willingly, although it remains unclear how this approach would translate into their actual research practice. It was the *Journal of Chemical Education* that decided to bring the two fields closer together by publishing a call for papers on the topic: "Reimagining Chemistry Education: Systems Thinking, and Green and Sustainable Chemistry."[21] The

---

20 IUPAC's website, project "Learning Objectives and Strategies for Infusing Systems Thinking into (Post)-Secondary General Chemistry Education," accessed 23/02/2022, https://iupac.org/projects/project-details/?project_nr=2017-010-1-050.
21 Peter G. Mahaffy, Edward J. Brush, Julie A. Haack, and Felix M. Ho, "Journal of Chemical Education Call for Paper Special Issue on Reimagining Chemistry Education: Systems Thinking, and Green and Sustainable Chemistry," *Journal of Chemical Education* 95 (2018): 1689–1691.

issue was published in 2019 with an eclectic range of articles trying to explain how systems thinking can help in conceptualising sustainability-related problems in the green chemistry education.

One of these rather theory-heavy articles explains the ambition of the endeavour in the following manner:

> Systems thinking identifies the interconnected components of a system and anticipates the behavior that results from the interaction of those components. Systems thinking connects structure (the components of the system) to behavior (system dynamics). ... Systems thinking requires thinking across spatial and temporal scales, and across disciplinary domains (molecular, biological, geological, economic, and social).
> [The project] calls upon educators to employ systems thinking to counter the reductionist approach ... and to prepare students to understand the 'molecular basis of sustainability.'[22]

The authors propose then an extremely large definition of green chemistry: "The practice of green chemistry stresses thoughtful design of molecules, materials, and processes to minimize adverse outcomes in humans and the environment through identification of the origin, transformation, and fate of atoms." Their article concludes with another strong statement:

> A systems thinking approach, epitomized by green chemistry, considers the unintended consequences of chemicals on human and environmental health. It also assesses the potential for the chemical enterprise to contribute to solving big global challenges, such as those outlined in the United Nations Sustainable Development Goals (UN SDGs). ... to ensure healthy lives, end poverty and hunger; ensure access to sanitation, affordable clean energy, and safe drinking water; grow strong and inclusive economies; reduce climate change and its impacts; and promote sustainable use of terrestrial and marine ecosystems.

Green chemistry is seen here as an epitome of the most holistic approach possible linking spheres of existence as separate as molecular, geological, economic, and social. It is presented as an essential tool for growing inclusive economies, ending hunger, and saving marine ecosystems. Twenty years prior, Graedel complained that green chemistry was overly narrow in its focus on synthesis and should operate on many levels, including societal. He made reference to systems thinking because the entire industrial ecology was rooted in this tradition. Did green chemistry make such a leap over the twenty years dividing both articles?

---

22 Katherine B. Aubrecht, Marie Bourgeois, Edward J. Brush, et al. "Integrating Green Chemistry in the Curriculum: Building Student Skills in Systems Thinking, Safety, and Sustainability," *Journal of Chemical Education* 96 (2019): 2872–2880.

Of course not. The 2019 article simply understands green chemistry in a different way than it was understood in 2001. Its authors equate it with any socially and environmentally friendly chemical endeavour and take for granted that green chemistry is a systems science. Why does it matter though? One could argue that the term simply evolved (as all terms do), and this evolution is happening in the direction that was envisaged by Graedel, or perhaps that scholars are free to define their frameworks as they please, and there are as many green chemistries as there are chemists willing to engage with the concept. The difficulty relates to the fact that the authors of the 2019 article still cite Anastas and Warner, their 12 principles, and the EPA's publications as the foundation of what they believe green chemistry is. They recast the knowledge generated in the disciplines as different as industrial ecology, ecotoxicology, and environmental chemistry into the green chemistry mould and fail to highlight this parentage. Instead, they make a strict connection to the EPA tradition as if the term "green chemistry" has always been understood in such an all-embracing way, which is simply not true.

In the entire debate over the place of systems thinking in green chemistry, it is easy to forget that all these discussions on integrating social, ethical, economic, and other dimensions into chemistry took place already half a century ago. In that period, it was not green but simply environmental chemistry that was the paradigmatic discipline of the new system-oriented mindset. Two major textbooks, both entitled *Environmental Chemistry*, from 1976 and 1978, can serve as an example.

In their 1976 book, John and Elizabeth Moore discuss a wide range of big themes: the history of chemical elements in the universe and their role in the emergence of life, the theoretical deliberations surrounding energy (including the problem of fossil fuels and new alternative energy sources), air pollution, water pollution and treatment, mineral scarcity (in the context of geology), and life sciences (with a focus on toxicity and food production).[23] The book concludes with a chapter "Science, Ethics, and Ecology" in which the authors advance the concept of "trans-scientific problems" that "transcend the limitation of scientific inquiry" and call for engagement of scientists in the solution of great challenges encountered by humanity. To reiterate: the origins of elements after the Big Bang and the social relevance of chemistry for providing food in developing areas are

---

23 John Moore and Elizabeth A. Moore, *Environmental Chemistry* (New York: Academic Press, 1976).

all discussed under the label of "environmental chemistry." It would be hard to find an even more all-embracing perspective.

The 1978 book edited by J.O'M. Bockris is a collection of chapters coming from different contributors and it is as eclectic as the previous one.[24] The book discusses overpopulation and the biochemical control of human fertility. It ponders potential chemical sources of synthetic food, the scarcity of mineral resources and fossil fuels, along with the development of alternative energies (especially hydrogen). One of the chapters on the environmental policy starts with a piece of advice:

> In order to function effectively environmental chemists must certainly have some concept of economics-not only the economics of textbooks, but also the economics of real life in which the pulse of progress is fueled by the motive of profit. It is necessary to see how the often restrictive demands of environmental chemistry interact with the present expensive free enterprise economic system.[25]

Environmental chemists are expected here to engage with the issues involving material flow from the social and political point of view.

The 1978 book concludes with a number of propositions for future lines of research in environmental chemistry such as energy conservation, more selective catalysis, mechanisms of selectivity in enzymes, photochemistry, and solution equilibria. All of this, twenty years before the 1998 green chemistry foundational text. The last big line of studies to explore for environmental chemists is described in the chapter "Research into Economics and Political Processes." Its author remarks somewhat angrily:

> Studies in economics and politics must grapple with gigantic problems arising from ecological considerations. The scientist brings the information. Should it be disseminated? How can it be disseminated without distortion? Understood? Absorbed? The relative significance understood in the face of so many clammerings for expenditure? How can the crowd be enlivened from the sodden mass of information-drunk blobs which it now is, directionless, hedonistic, influenced only by the groups who have money to pay the media to print the loaded message, making more offerings to the crowd's hedonism? ... It is absolutely clear that in the study and practice of politics and economics lie many of the answers to the problems of environmental chemistry.[26]

---

24 John O'M. Bockris, ed., *Environmental Chemistry* (New York: Plenum Press, 1978).
25 Burton H. Klein, "The Public Policy Issues Involved in Dealing with Environmental Degradation: A Dynamic Approach," in John O'M. Bockris, ed., *Environmental Chemistry* (New York: Plenum Press, 1978), 749–769.
26 Klein, "The Public Policy Issues Involved in Dealing with Environmental Degradation," 769.

It is from these early works in politically engaged, socially oriented, and systems-aware environmental chemistry that evolved modern fields of inquiry such as sustainability studies, industrial ecology, bioeconomy, green chemistry and many more. Over the following ten years, in the late 1970s and in the 1980s, environmental chemists' scope of activities narrowed down to the study of environmental pollution. This focus helped in forging new more restrictive environmental regulations, which put pressure on the chemical industry to limit toxic discharges. And yet it is against this approach that revolted early green chemistry pioneers who envisaged working on pollution prevention on the molecular level, meanwhile opting for a less politically engaged, less overarching, but more practically focused and industry-friendly language.

By incorporating systems thinking into green chemistry, as envisaged by the *Journal of Chemical Education* in 2019, and by transforming it into an all-embracing eco-friendly idea on how to deal with global challenges of food hunger and maritime pollution, we make a full circle. Green chemistry becomes then, from the theoretical point of view, the same thing environmental chemistry aspired to be in the 1970s, even if less explicitly political. This is not yet the case from the practical point of view. The *Green Chemistry* journal remains relatively conservative and systems thinking is absent from its pages. However, in a less constrained environment, such as the *Journal of Chemical Education*, these ideas circulate and may influence the way the concept is understood. But are we actually better equipped to deal with these challenges today than fifty years ago? Why did environmental chemistry of the 1970s fail to deliver on its promises? And how is it possible that this parentage was forgotten? These are some problems that will unavoidably rise if we want to develop green chemistry in new directions.

2.2. Green chemistry and social justice

Solving global challenges in a systemic manner, as envisaged by the 1970s books in environmental chemistry, was not just about dispassionate analysis. It was also about political activism and taking a stance. Green chemistry was never truly apolitical either. On the one hand, we have seen a strong distaste of some of its proponents for regulation as well as their pro-business language conditioning the greening of chemistry on economic profitability. On the other hand, green chemistry attracted the attention of left-leaning environmentalists, in particular from social sciences, from very early on as well. They often jumped on the bandwagon of green chemistry believing they could contribute to it and extend its scope. I have already discussed the involvement of the political scientist Ted

Woodhouse who fought for additional funding for green chemistry in US Congress and who even contributed to the development of the field conceptually in its early years with the famous brown/green divide.

This latter direction was not appreciated by everyone. In 2017, *Foundations of Chemistry*, the world's leading journal in philosophy of chemistry, published a short monograph, a memoir of sorts, of a former industrial chemist Mark A. Murphy under the title "Early Industrial Roots of Green Chemistry and the history of the BHC Ibuprofen process invention and its Quality connection."[27] In his article, Murphy argues that much of what the EPA's green chemists advocated in the 1990s had been already practised by the industry in the 1980s and offers a vivid recollection of some of the key technological advancements in the field. However, he also shares his thoughts on more recent trends and takes a very negative stance towards the social science shift in green chemistry by describing his experiences from a green chemistry conference in 2012 organized by the American Chemical Society.

> I attended a conference session on "Social Science Perspectives on Green Chemistry." An academic speaker from a prominent business school reported, clearly approvingly, significant evidence of a growing "collective identity" sub-movement within the Green Chemistry "Movement." When the presentation was over, I asked a question, namely what was the benefit of a "collective identity" sub-movement within Green Chemistry "movement'?' Although the speaker's response went on for quite a while, and I have a Ph.D. in the sciences, and a Law degree, and practical experience in both fields, I literally couldn't understand the answer I received.

Murphy then continues by criticizing the speech of an environmental activist during the same event.

> At a break in the session, in the hall, I approached the "activist" speaker hoping for a private, rational conversation. I started to express my concerns, but within less than a minute, the "activist" speaker began very loudly, angrily, and repeatedly denouncing me personally (even though she had never previously known of me) as a corrupt and incompetent industry lawyer/stooge!![28]

Murphy expresses his objections to the role of social science in shaping green chemistry's identity and even more to environmental activism. He attacks

---

27 Mark A. Murphy, "Early Industrial Roots of Green Chemistry and the history of the BHC Ibuprofen process invention and its Quality connection," *Foundations of Chemistry* 20 (2018): 121–165.
28 Murphy, "Early Industrial Roots of Green Chemistry and the history of the BHC Ibuprofen process invention and its Quality connection," 158.

Woodhouse and Breyman's 2005 article in which they argued against leaving the ecological transformation in the hands of technoscientists. Murphy responds that he is much less confident in social science academic professors than in his colleagues from the chemical industry.

Whether one agrees with Murphy's political position or not, it is important to point out that there is a genuine question of how social and political science, or environmental activism for that matter, can contribute to green chemistry in practice.

Some answers can be found in the yearly Green Chemistry and Engineering Conference, which featured for the last few years a dynamically growing symposium on green chemistry, social justice, and equity. Why was the symposium getting increasingly popular? Two American scholars, Grace A. Lasker and Edward Brush, offered their insights on the issue in the article "Integrating social and environmental justice into the chemistry classroom: a chemist's toolbox" published in 2019 in *Green Chemistry Letters and Reviews*:

> The growth of this symposium is not surprising. Chemistry professionals connect with green chemistry from science, education, and business perspectives. Others intersect from areas such as public health, toxicology, occupational health, policy, government, engineering, industry, etc. Linking green and sustainable chemistry to equity and environmental justice inspires and resonates with a completely new audience and has expanded the reach of green chemistry's influence.[29]

Lasker and Brush advocate in their paper in favour of a more prominent integration of social justice themes in the green chemistry teaching:

> Although the chemical enterprise has provided numerous contributions to humanity, unintended consequences contribute to a disproportionate exposure of hazardous chemicals to certain populations based on race and socioeconomic status. Integrating concepts of social and environmental justice within chemistry curriculum provides an educational framework to help mitigate these impacts by training the next generation of chemists with justice-centered and green chemistry principles to guide their future work.

The authors carefully avoid the term "social class," but the core element of their ambition remains relatively clear. Blue-collar workers and people from low-income households, in the US often from ethnic minorities, are usually the first victims of exposure to dangerous chemicals, be it in accidents, through a

---

29 Grace A. Lasker and Edward J. Brush, "Integrating social and environmental justice into the chemistry classroom: a chemist's toolbox," *Green Chemistry Letters and Reviews* 12 (2019): 168–177.

calculated risk of their professions, or as a result of more subtle forms of discrimination such as housing policy. To remedy these issues, the authors call for teaching green and justice-centred chemistry. Already at this stage, one could ask whether the authors do not expect too much from green chemistry, if we understand it as molecular pollution prevention, and whether it can really address the consequences of decades of systemic racism and gargantuan inequalities ravaging the US, but the caveat is the term 'justice-centred,' which suggests that there is more to be taken into account than narrowly construed green chemistry.

Things get, however, more problematic in the following passage:

> Since chemicals know no borders, it is important that students recognize the global impact of the work they will be doing and that they grow a social consciousness that allows for green chemistry and justice principles to guide pre-production design decisions. In the classroom, faculty could facilitate having weekly discussions around case studies that highlight regrettable substitutions, environmental, or human health impacts of chemicals such as DDT, lead, asbestos, chemicals in cigarettes, eWaste, noenicitinoid and chlorinated pesticides, etc.

These are all important issues, some in the spotlight since the 1960s, but one does not need the green chemistry framework to discuss any of them. The problem of lead pollution, disquieting environmental consequences of asbestos, and especially DDT has been at the heart of the preoccupations of environmental chemistry for years. It appears, again, that the authors understand green chemistry in an extremely generous manner, as if it was some sort of broad environmental paradigm, just like the paper I analysed in the previous section that conflated green chemistry and systems science.

The problems surrounding the definition and purpose of green chemistry are even more evident in the following part of the article focusing on the pedagogy of chemistry.

> It is important that the space that chemistry exists within (be that the classroom, laboratory, office, program, community, etc.) be inclusive, diverse, and welcoming to all students but particularly mindful toward the additional work required to support underrepresented minority students. One approach is applying critical race pedagogy in the classroom. Critical race pedagogy 'is an instructional approach designed to challenge and transform the prevailing Eurocentric power structure that organizes higher education curricula in order to cultivate spaces that validate the experiences of Students of Color.'

The German reviewer of the 1993 sanfte Chemie book who argued that it undermined the achievements of Western Civilization would have probably his fears confirmed seeing that modern green chemistry educators wish to challenge the

Eurocentric power structures! Still, even if we fully agree with the authors' assessment, the question still stands: why should these problems appear specifically in the context of green chemistry (the article was after all published in *Green Chemistry Letters and Reviews*)?

This and other articles exploring the problem of social and environmental justice in the context of chemistry education treat green chemistry as if it was an entry point for students to discuss a very broad range of politically charged socio-environmental issues. However, following this logic, green chemistry classes would then become a substitute for a proper education in STS (science and technology studies/social studies of science), scientific culture, environmental history, and sustainability, sprinkled with elements of environmentally friendly chemistry. This, of course, is not the green chemistry as it is understood in research journals or through any number of green principles. The social justice-minded partisans of green chemistry were at least somewhat aware of the limitations of the field traditionally construed and discussed incorporating additional *"justice principles"* to the 12 principles of green chemistry, to extend its scope.[30] As of 2022, no such formalisation appears to have taken place, but if it happens, we would have to rethink again the limits of the concept.

In fact, does the concept have any boundaries at all if it stretches, in the spirit of systems thinking, from environmental justice to empirical analyses of ionic solvents? And how to even meaningfully engage with the framework so inclusive? If green chemistry develops along the lines of systems thinking and social justice, the discipline's identity will be certainly called into question in the following years.

Conclusions

The frontiers of green and sustainable thinking in chemistry are not set in stone. New ideas continue to emerge incessantly. Should we move towards a circular economy by adopting the new circular chemistry model? Or rather operate within previous frameworks reinterpreting them appropriately? Should green chemistry incorporate more explicitly elements of systems thinking? But how would it differ then from the environmental chemistry of yore? Or maybe we should keep green chemistry as it is, but engage with a higher-level concept of

---

30 Grace A. Lasker, Karolina E. Mellor, Melissa L. Mullins, et al. "Social and Environmental Justice in the Chemistry Classroom," *Journal of Chemical Education* 94 (2017): 983–987.

one-world chemistry instead? And what about the political dimension of green chemistry? After all, the environmental chemistry of the 1970s was militant in its criticism of the chemical industry. Perhaps the new green chemistry could become more politically involved too and contribute to our fight against exclusion and structural forms of injustice. But then, by incorporating too much politics into green or sustainable chemistry, do we not risk that these concepts become abused by political advocacy groups whose interpretation of what is sustainable is unconventional at best (or simply contrary to what we envisage as sustainable)?

All these questions can be seen in two opposite ways. On the one hand, they can be understood as a vivid and intellectually stimulating debate about the frontiers of sustainability, thus paving the way into a better future. After all, these debates are not a goal in themselves, they are to change something, to make things right, to better address the challenges we face. On the other hand, I believe a much more bleak interpretation is possible. The core problems discussed today are not substantially different from the ones already signalled in the 1960s and 1970s. We have not moved forward on many issues since then. How have we arrived at such a deplorable state of affairs? It appears that well-meaning scholars formulate novel concepts without understanding or even familiarising themselves with the ones developed in the past. In fact, the horizon of past papers that are being cited in new works is becoming shorter and shorter, different traditions interact with each other in a superficial way, and people keep reinventing the wheel. No overarching view of these problems is being constructed and, most importantly, no critical evaluation of new, or old for that matter, concepts is ever formulated. Virtually all the programmatic papers I cited throughout this book are very enthusiastic about new ideas. High-profile journals publish willingly articles laying down new fancy theoretical frameworks, but almost never papers that would criticize and dismiss them. Perhaps we should rethink whether we genuinely need a new circular chemistry or why exactly the introduction of systems thinking is useful in green chemistry classes.

And yet scientists may not be the ones to blame. Many researchers have not even enough time to read all empirical papers relevant to their fields of study, and even fewer to dig through dozens of theoretical, conceptual, and programmatic articles and chapters published over the last few decades. On top of that, in the era of publish or perish, scientists are almost encouraged to advance half-baked ideas full of buzzwords, just to gain attention, press coverage, social media presence, and citations. 'Slow' and critical thinking is not rewarded. This systemic neglect in engaging in a meaningful discussion over new trends in science may explain why the debates on environmental chemistry in the early 2020s are so similar to the ones from the 1970s, and why the future does not necessarily look as green as we would hope.

General conclusions: green chemistry as history of scientific ideas

Green chemistry, sustainable chemistry, circular chemistry, soft chemistry, and one-world chemistry are all names given to an incessant attempt to go beyond the boundaries of traditionally construed chemistry and engage, in some way, with what scientists throughout the consecutive decades believed to be the most pressing environmental and societal issues. There has been an underlying assumption behind all these philosophies that chemistry is somehow insufficient for the modern world, that it creates challenges it is incapable of overcoming, and that it needs a radical change. These various concepts tried to break the stalemate and show that chemistry could become more relevant for the challenges humanity is facing.

To be sure, someone more sympathetic to science could wonder whether it is really chemistry that is to blame for the Bhopal disaster, for the cancer epidemics, or for a whole range of environmental injustices. Aren't chemists taking blame here for structural deficiencies of our economic system, political misjudgements, and social problems that are far beyond what they actually can change as chemistry professionals? A sceptic could go in their criticism even further: isn't green (and sustainable) chemistry counterproductive by claiming that it can address these issues? Perhaps green chemistry became a tool for presenting inherently political and economic problems as solvable through technical solutions? I do not pretend to be sure of any answers, but I strongly believe these questions deserve to be asked by chemists and non-chemists alike. However, before engaging in the discussion, the elementary terms need to be laid down and explained in their historical context.

One of the most interesting problems I tried to convey throughout this work is that despite the tremendous rise in the popularity of concepts of green and sustainable chemistry, their use in the literature has not become more consistent. On the contrary, not only do experimental scientists often practise green chemistry based on 'gut feeling' about what feels green or not, but even scholars who actually engage with theoretical, historical, and programmatic problems often remain unfamiliar with what the field has already accomplished. One can wonder whether there is any coherent methodology at all across different publications referring to these various frameworks and whether there is any connection between the theory of green chemistry and the practice of green chemistry.

The point is not to prove that the various conceptualisations of greenness formulated over the last thirty years are useless. The question is rather whether there is any way to turn the fragmented bits of theory, competing frameworks, eclectic case studies on fashionable topics, and intuitive ideas about greenness into a coherent and helpful way of thinking for chemistry students, chemists in academia, researchers in the private sector, and policymakers alike. Is there a way to make all these people operate a single mutually intelligible language? Or to increase the stakes: is there any way to prevent green (and sustainable) chemistry from collapsing under its own weight, accumulated over the years of expansion? After all, when the concepts become too general and too vague, they also turn useless.

What I argue for is that we should turn green chemistry into a 'historical' science. Or to put it differently, we should take a step back and think about it and teach about it through the lenses of the history of science, technology, knowledge, and especially of the history of ideas. The goal would be to probe into interconnections of (proto/meta)scientific ideas concerning the relationship between human beings and the environment using the tools of history and other social sciences. This historical green chemistry would then become the study of conceptual frameworks produced by environmentally-minded chemists over decades. It would include the reflection on chemurgy, and on the reasons for its ultimate failure, on the works and ways of thinking of the environmental chemists in the 1970s, and on the reaction to the perceived insufficiencies of these older approaches to address the challenges we label as sustainability-related today. Such historical green chemistry would discuss the reasons behind the development of different sets of guiding principles and evaluate their impact. It would bring together codifications such as the 12 principles of green chemistry, the 12 principles of green engineering, Christ's 8 rules for the chemical industry, the 12 principles of circular chemistry, the 15 key characteristics of sustainable chemistry, and many others, in order to thoroughly engage with all of them, present their strengths and weaknesses, explain similarities and differences, and most importantly reflect on their failures and successes. Why were they formulated with what purpose? Were they descriptive, normative, or somewhere in between? How did they influence the wider community of chemists?

The historical green chemistry could at the same discuss the evolution of 'greenness', 'sustainability', and 'environmental friendliness' criteria, providing an appropriate economic, political, and scientific context. The purpose of this approach would be to highlight common threads; ideas shared by different communities and developed sometimes independently over different periods. All of this to facilitate their translation. An encounter between environmental chemists working food supply in the 1970s, industrial ecology scholars studying local closed feedstock cycles in the 1990's, and one-world chemistry proponents advocating systems thinking in the 2020s could potentially open wide new horizons. Only after this preliminary task, we would be able to probe more in detail into empirical results generated over the decades in different disciplines and reflect on whether they genuinely contributed to making our society and economy more sustainable and environmentally friendly. The historical green chemistry would act as a roadmap for those who aspire to better understand chemistry's contributions in these areas. It would not try to replace the existing frameworks with new ones, but rather to accommodate them in a historical narrative. Such an approach would help to avoid the phenomenon of reinventing the wheel and may help in constructing more robust, better empirically informed ways for making chemistry more sustainable and our future greener. I am strongly convinced that this book is the first step in this direction.

Bibliography

Abraham, Martin A. and Nhan Nguyen. "Green engineering: Defining the principles," *Environmental Progress* 22 (2004): 233–236.

Allen, David, Julie Carrier, Jingwen Chen et al. "The Evolution of ACS Sustainable Chemistry & Engineering," *ACS Sustainable Chemistry and Engineering* 8 (2020): 1–1.

Allen, David. "Welcome to ACS Sustainable Chemistry & Engineering," *ACS Sustainable Chemistry and Engineering* 1 (2013): 1.

Alonso, David Martin, Jesse Q. Bond and James A. Dumesic. "Catalytic conversion of biomass to biofuels," *Green Chemistry* 12 (2010): 1493–1513.

Amato, Ivan. "The Slow Birth of Green Chemistry," *Science* 259 (1993): 1538–1541.

Anastas, Paul and Carol A. Farris, ed. *Benign by Design: Alternative Synthetic Design for Pollution Prevention* (Washington, D.C.: ACS Symposium Series, 1994).

Anastas, Paul and Evan S. Beach. "Green chemistry: the emergence of a transformative framework," *Green Chemistry Letters and Reviews* 1 (2007): 9–24.

Anastas, Paul and John C. Warner. *Green Chemistry: Theory and Practice* (Oxford: Oxford University Press, 1998).

Anastas, Paul and Joseph Breen. "Design for the environment and Green Chemistry: the heart and soul of industrial ecology," *Journal of Cleaner Production* 5 (1997): 97–102.

Anastas, Paul and Julie Zimmerman. "Design through the 12 Principles of Green Engineering," *Environmental Science and Technology* 37 (2003): 94A–101A.

Anastas, Paul and Mary Kirchhoff. "Origins, current status, and future challenges of green chemistry," *Accounts of Chemical Research* 25, n. 9 (2002): 686–694.

Anastas, Paul and Nicolas Eghbali. "Green Chemistry: Principles and Practice," *Chemical Society Reviews* 39, No. 1 (2010): 301–312.

Anastas, Paul and Roger Lankey. "Life cycle assessment and green chemistry: the yin and yang of industrial ecology," *Green Chemistry* 2 (2000): 289–295.

Anastas, Paul and Tracy C. Williamson, ed. *Green Chemistry. Designing Chemistry for the Environment* (Washington D.C.: American Chemical Society, 1996).

Anastas, Paul and Tracy C. Williamson. "Green Chemistry: An Overview," in *Green Chemistry. Designing Chemistry for the Environment*, ed. Paul Anastas and Tracy C. Williamson (Washington D.C.: American Chemical Society, 1996), 1–17.

Anastas, Paul, and Mary M. Kirchhoff. "Origins, Current Status, and Future Challenges of Green Chemistry," *Accounts of Chemical Research* 35 (2002): 686–694.

Anastas, Paul, and Nicolas Eghbali. "Green Chemistry: Principles and Practice," *Chemical Society Reviews* 39 (2010): 301–312.

Anastas, Paul, Buxing Han, Walter Leitner, Walter Leitner and Martyn Poliakoff. "'Happy silver anniversary:' Green Chemistry at 25," *Green Chemistry* 18 (2016): 12–13.

Anastas, Paul, David StC. Black, Joseph Breen, Terrence Collins, Sofia Memoli, Junshi Miyamoto, Martyn Polyakoff, and William Tumas. "Synthetic pathways and processes in green chemistry. Introductory overview," *Pure and Applied Chemistry* 72 (2000): 1207–1228.

Anastas, Paul. "Benign by Design Chemistry," in *Benign by Design: Alternative Synthetic Design for Pollution Prevention*, ed. Paul Anastas and Carol A. Farris (Washington, D.C.: ACS Symposium Series, 1994), 2–22.

Anastas, Paul. "Joe Breen – heart and soul of Green Chemistry," *Green Chemistry* 1 (1999): G87.

Anastas, Paul. "Origins and Early History of Green Chemistry," in *Advanced Green Chemistry. Part 1: Greener Organic Reactions and Processes*, ed. I. T Horváth and M. Malacria (London: World Scientific, 2018): 1–17.

Andraos, John. "Useful Tools for the Next Quarter Century of Green Chemistry Practice – A Dictionary of Terms and a Dataset of Parameters for High Value Industrial Commodity Chemicals," *ACS Sustainable Chemistry and Engineering* 6 (2018): 3206–3214.

Angrick, Michael, Klaus Kümmerer and Lothar Meinzer. *Nachhaltige Chemie. Erfahrungen und Perspektiven* (Marburg: Metropolis, 2006).

Antonietti, Markus and Maria-Magdalena Titirici. "Coal from carbohydrates: The 'chimie douce' of carbon," *Comptes Rendus Chimie* 13 (2010): 167–173.

Armstrong, Laura B., Mariana C. Rivas, Michelle C. Douskey, Anne M. Baranger. "Teaching Students the Complexity of Green Chemistry and Assessing Growth in Attitudes and Understanding," *Current Opinion in Green and Sustainable Chemistry* 13 (2018): 61–67.

Artz, Jens, Thomas E. Müller, Katharina Thenert, Johanna Kleinekorte, Raoul Meys, André Sternberg, André Bardow, and Walter Leitner. "Sustainable Conversion of Carbon Dioxide: An Integrated Review of Catalysis and Life Cycle Assessment," *Chemical Reviews* 118 (2018): 434–504.

Asfaw, Nigist, Yonas Chebude, Andinet Ejigu, Bitu B. Hurisso, Peter Licence, Richard L. Smith, Samantha L. Y. Tang and Martyn Poliakoff. "The 13

Principles of Green Chemistry and Engineering for a Greener Africa," *Green Chemistry* 13 (2011): 1059–1060.

Aubrecht, Katherine B., Marie Bourgeois, Edward J. Brush, Jennifer Mackellar, Jane E. Wissinger. "Integrating Green Chemistry in the Curriculum: Building Student Skills in Systems Thinking, Safety, and Sustainability," *Journal of Chemical Education* 96 (2019): 2872–2880.

Augé, Jacques and Marie-Christine Scherrmann. *Chimie verte: Concepts et applications* (Paris: CNRS Editions, 2017).

Ayres, Robert U. "Industrial Metabolism" in *Technology and Environment*, ed. Jesse H. Ausubel and Hedy E. Sladovich (Washington D.C.: National Academy Press, 1989): 23–49.

Bashkin, James K. "Green Chemistry: Can we rally together, or will we fragment into pieces?," *Green Chemistry* 4 (2002): G14.

Becher, D. "Vermeiden, Vermindern, Verwerten – integrierter Umweltschutz in der Produktion," in the report *Die Bayer-Umweltperspektive II* (Leverkusen, 1991).

Bianucci, Giovanni and Esther Ribaldone Bianucci. *Il trattamento delle acque residue industriali e agricole* (Milano: Hoepli, 1992).

Blum, Christopher, Dirk Bunke, Maximilian Hungsberg Elsbeth Roelofs, Anke Joas, Reinhard Joas, Markus Blepp, Hans-ChristianStolzenberg. "Das Konzept der Nachhaltigen Chemie: Schlüsselfaktoren für den Übergang zu einer nachhaltigen Entwicklung," *Sustainable Chemistry and Pharmacy* 13 (2019): 100140.

Bockris, John O'M. ed., *Environmental Chemistry* (New York: Plenum Press, 1978).

Boehlert, Sherwood. Report prepared for the Green Chemistry Research and Development Act of 2005, report No. 109–82, 2005.

Bornscheuer, Uwe, Gjalt Huisman, Romas Kazlauskas, S. Lutz, J. C. Moore and K. Robins. "Engineering the third wave of biocatalysis," *Nature* 485 (2012): 185–194.

Böschen, Stefan, Dieter Lenoir, and Martin Scheringer. "Sustainable chemistry: starting points and prospects," *Naturwissenschaften* 90 (2003): 93–102.

Bourg, Dominique and [Suren Erkman, ed. *Perspectives on Industrial Ecology* (New York: Greenleaf Publishing, 2003).

Bozell, Joseph J. and Gene R. Petersen. "Technology development for the production of biobased products from biorefinery carbohydrates," *Green Chemistry* 12 (2010): 539–554.

Breen Joseph J. and Michael. J. Dellarco. *Pollution Prevention in Industrial Processes: The Role of Process Analytical Chemistry* (Washington D.C.: American Chemical Society, 1992).

Brindle, Ian D. "Green chemistry—a Canadian perspective," *Green Chemistry* 1 (1999): 155–157.

Bringezu, Stefan. "Industrial Ecology and Material Flow Analysis" in *Perspectives on Industrial Ecology*, Greenleaf Publishing, ed. Dominique Bourg and Suren Erkman (London: Routledge, 2003).

Burgman, Mark, Mike Tennant, Nikolaos Voulvoulis, Karen Makuch and Kaveh Madani. "Facilitating the transition to sustainable green chemistry," *Current Opinion in Green and Sustainable Chemistry* 13 (2018): 130–136.

Cahn, Robert W. *The Coming of Materials Science* (Londres: Pergamon, 2001).

Caradonna, Jeremy L. *Routledge Handbook of the History of Sustainability* (New York: Routledge, 2018).

Caradonna, Jeremy L. *Sustainability: A History* (Oxford: Oxford University Press, 2014).

Carolan, Michael S. "Ethanol versus Gasoline: The Contestation and Closure of a Socio-technical System in the USA," *Social Studies of Science* 39 (2009): 421–448.

Carra, Joseph. "International Diffusion of Sustainable Chemistry," *Proceedings of the OECD workshop on sustainable chemistry, Series on Risk Management OECD* (Paris: OECD, 1999): 47–50.

Centi, Gabriele and Siglinda Perathoner. "From Green to Sustainable Industrial Chemistry" in *Sustainable Industrial Processes*, ed. Fabrizio Cavani, [Gabriele Centi, [and Siglinda Perathoner (Weinheim: Wiley-VCH, 2009), 1–69.

Chang, Kenneth. "Yves Chauvin, 'Green' Chemist and Nobel Laureate, Dies at 84," *The New York Times*, 30 January 2015, https://www.nytimes.com/2015/01/31/science/yves-chauvin-chemist-sharing-nobel-prize-dies-at-84.html

Chatel, Gregory. "Chemists around the World, Take Your Part in the Circular Economy!," *Chemistry. A European Journal* 26 (2020): 9665–9673.

Chemat, Farid, Maryline Abert Vian, and Giancarlo Cravotto. "Green extraction of natural products: concept and principles," International Journal of Molecular Sciences 13 (2012): 8615–8627.

Chen, Tse-Lun, Hyunook Kim, Shu-Yuan Pan, Po-Chih Tseng, Yi-Pin Lin, Pen-Chi Chiang. "Implementation of green chemistry principles in circular economy system towards sustainable development goals: Challenges and perspectives," *Science of the Total Environment* (10 May 2020), 716: 136998.

Chevalier, Jacques. "Le carburant national en Allemagne," *Bois et Résineux*, 30 April, 1922, 1.

Chheda, Juben N., George W. Huber, and James A. Dumesic. "Liquid-phase catalytic processing of biomass-derived oxygenated hydrocarbons to fuels and chemicals," *Angewandte Chemie – International Edition* 46 (2007): 7164–7183.

Choi, Sunho, Jeffrey H. Drese, and Christopher W. Jones. "Adsorbent materials for carbon dioxide capture from large anthropogenic point sources," *ChemSusChem* 2 (2009): 796–854.

Christ, Claus. "Umweltschutz in der chemischen Industrie – Vermindern und vermeiden von Abfallen," in *Umwelt, Logistik und Verkehr*, ed. Reinhardt Junemann (Dortmund: Praxiswissen GmbH, 1992).

Christ, Claus. *Production-Integrated Environmental Protection and Waste Management in the Chemical Industry* (Weinheim: Wiley-VCH, 1999).

Christie, Ian, Richard M. Gunton and Adam P. Hejnowicz. "Sustainability and the common good: Catholic Social Teaching and 'Integral Ecology' as contributions to a framework of social values for sustainability transitions," *Sustainability Science* 14 (2019): 1343–1354.

Clark, James and Duncan Macquarrie. *Handbook of Green Chemistry and Technology* (Oxford: Blackwell Science, 2002).

Clark, James. "Green chemistry initiatives in Japan," *Green chemistry* 4 (2002): G54.

Clark, James, Roger Sheldon, Colin Raston, Martyn Poliakoff and Walter Leitner. "15 years of Green Chemistry," *Green Chemistry* 16 (2014): 18.

Clark, James. "Green biorefinery technologies based on waste biomass," *Green Chemistry* 21 (2019): 1168–1170.

Clark, James. "Green chemistry: challenges and opportunities," *Green Chemistry* 1 (1999): 1–8.

Clark, James. "Green chemistry: today (and tomorrow)," *Green Chemistry* 8 (2006): 17–21.

Coelho, Maria Alice and Bernardo D. Ribeiro, *White Biotechnology for Sustainable Chemistry*, (Cambridge: RSC Green Chemistry, 2016).

Collins, Terrence. "Introducing Green Chemistry in Teaching and Research," *Journal of Chemical Education* 72 (1995): 965–966.

Collins, Terrence. "A call for a chlorine sunset," *Nature* 406 (2000): 17–18.

Collins, Terrence. "Towards Sustainable Chemistry," *Science* 291 (2001): 48–49.

Colonna, Paul, ed. *Chimie Verte* (Paris: Technique Et Documentation, 2006).

Constable, David J. C. "What Do Patents Tell Us about the Implementation of Green and Sustainable Chemistry?," *ACS Sustainable Chemistry and Engineering* 8 (2020): 14657–14667.

Constable, David J.C. "The practice of chemistry still needs to change," *Current Opinion in Green and Sustainable Chemistry* 7 (2017): 60–62.

Constable, David J.C. and Concepción Jiménez- González. *Green Metrics* (Weinheim: Wiley-VCH, 2018).

Curzons, Alan D., David J. C. Constable, David N. Mortimera and Virginia L. Cunningham. "So you think your process is green, how do you know?—Using principles of sustainability to determine what is green–a corporate perspective," *Green Chemistry* 3 (2001): 1–6.

Dadabhoy, Anjum and Peter Gölitz. "Sustainability: Chemistry Is Key," *ChemSusChem* 1 (2008): 4–5.

Deborah L. Illman. "'Green' technology presents challenge to chemists," *Chemical & Engineering News* 71 (1993): 26–30.

Deetlefs, Maggel and Kenneth R. Seddon. "Assessing the greenness of some typical laboratory ionic liquid preparations," *Green Chemistry* 12 (2010): 17–30.

DeVito, Stephen C. and Roger L. Garrett. *Designing Safer Chemicals Green Chemistry for Pollution Prevention* (Washington D.C.: American Chemical Society, 1996).

Dicks, Andrew P. and Andrei Hent. *Green Chemistry Metrics. A Guide to Determining and Evaluating Process Greenness* (New York: Springer, 2014).

Ducamp, Christine. "Chimie verte: approche nouvelle et responsable face aux problèmes issus des activités chimiques," in *Développement durable et autres questions d'actualité – Questions socialement vives dans l'enseignement et la formation*, ed. Alain Legardez (Dijon: Educagri, 2011), 145–162.

Eastes, Richard-Emmanuel and Bernadette Bensaude-Vincent, ed. *Philosophie de la chimie* (Paris: De Boeck Superieure).

Ehrenfeld, John R. "Industrial Ecology and LCM: Chicken and Egg?," *The International Journal of Life Cycle Assessment* 8 (2003): 59–60.

Ekerholm, Helena. "Cultural Meanings of Wood Gas Automobile Fuel in Sweden, 1930–1945," in *Past and present energy societies: How energy connects politics, technologies and cultures*, ed. Nina Möllers and Karin Zachmann (Bielefeld: Transcript Verlag, 2012), 223–247.

Erythropel, Hanno C., Julie B. Zimmerman, Tamara M. de Winter, Laurène Petitjean, Fjodor Melnikov, Chun Ho Lam, Amanda W. Lounsbury, Karolina E. Mellor, Nina Z. Janković, Qingshi Tu, Lauren N. Pincus, Mark M. Falinski, Wenbo Shi, Philip Coish, Desirée L. Plata and Paul T. Anastas. "The Green

ChemisTREE: 20 years after taking root with 1 the 12 Principles," *Green Chemistry* 20 (2018): 1929–1961.

Finlay, Mark R. "Old Efforts at New Uses: A Brief History of Chemurgy and the American Search for Biobased Materials," *Journal of Industrial Ecology* 7 (2003): 33–43.

Fischer, Hermann and Horst G. Appelhagen. *Chemiewende: Von der intelligenten Nutzung natürlicher Rohstoffe* (Antje Kunstmann, 2017).

Fischer, Hermann. *Plädoyer für eine Sanfte Chemie* (Braunsweig: Allembik Verlag, 1993).

Friege, Henning. "Sustainable Chemistry – A concept with important links to waste management," *Sustainable Chemistry and Pharmacy* 6 (2017): 57–60.

Frosch, Robert A. and Nicholas E. Gallopoulos. "Strategies for Manufacturing," *Scientific American* 261 (1989): 144–152.

Garnier, Estelle. "Une approche socio-économique de l'orientation des projets de recherche en chimie doublement verte" (PhD diss., University Reims Champagne-Ardenne, 2012).

Garrett, Roger L. "Pollution Prevention, Green Chemistry, and the Design of Safer Chemicals," in *Designing Safer Chemicals Green Chemistry for Pollution Prevention*, ed. Stephen C. DeVito and Roger L. Garrett (Washington D.C.: American Chemical Society, 1996), 2–15.

Gilles, Laure and Sylvain Antoniotti. "Chimie durable et parfumerie," *Nez, la revue olfactive* (2020). https://mag.bynez.com/science/chimie-durable-et-parfumerie/.

Girard, Simon A., Thomas Knauber, Chao-Jun Li. "The cross-dehydrogenative coupling of C(sp3)-H bonds: a versatile strategy for C-C bond formations," *Angewandte Chemie – International Edition* 53 (2014): 74–100.

Gleich, Arnim von. *Der wissenschafltiche Umgang mit Natur – Über die Vielfalt harter und sanfter Naturwisseschaften* (Frankfurt: Campus Verlag, 1989).

Graedel, Thomas. "Green chemistry as systems science," *Pure and Applied Chemistry* 73 (2001): 1243–1246.

Graedel, Thomas. "Green chemistry in an industrial ecology context," *Green Chemistry* 1 (1999): G126–G128.

Green, Sarah A. "Progress and Barriers," in *Sustainable Green Chemistry*, ed. Mark A. Benvenuto (Berlin: Walter de Gruyter, 2017): 17–28.

Grober, Ulrich. *Sustainability: A Cultural History* (Totnes: Green Books, 2012).

Guo, Hui-Lin, Xian-Fei Wang, Qing-Yun Qian, Feng-Bin Wang, Xing-Hua Xia. "A Green Approach to the Synthesis of Graphene Nanosheets," *ACS Nano* 3 (2009): 2653–2659.

Hall, Nina. "Chemists clean up synthesis with one-pot reactions," *Science*, 266 (1994): 32–34.

Hancock, Kenneth G. and Margaret A. Cavanaugh, "Environmentally Benign Chemical Synthesis and Processing for the Economy and the Environment," in *Benign by Design: Alternative Synthetic Design for Pollution Prevention*, ed. Paul Anastas and Carol A. Farris (Washington, D.C.: ACS Symposium Series, 1994), 23–30.

Hofer, Rainer, ed. *Sustainable Solutions for Modern Economies* (Cambridge: RCS Publishing, 2009).

Hoffman, Andrew. *From Heresy to Dogma: An Institutional History of Corporate Environmentalism* (San Francisco: New Lexington Press, 1997).

Homburg, Ernst and Elisabeth Vaupel. *Hazardous Chemicals: Agents of Risk and Change, 1800–2000* (New York: Berghahn Books, 2019).

Horváth, István T. and Paul Anastas. "Innovations and green chemistry," *Chemical Reviews*, 107, No. 6 (2007): 2169–2173.

Huddleston, Jonathan G., Ann E. Visser, W. Matthew Reichert, Heather D. Willauer, Grant A. Broker, and Robin D. Rogers. "Characterization and comparison of hydrophilic and hydrophobic room temperature ionic liquids incorporating the imidazolium cation," *Green Chemistry* 3 (2001): 156–164.

Hutzinger, Otto. "The Greening of Chemistry- Is It Sustainable?" *Environmental Science and Pollution Research* 6 (1999): 123.

Ida, Hisao. "The Japanese approach to Sustainable Chemistry," *Proceedings of the OECD workshop on sustainable chemistry, Series on Risk Management OECD* (Paris: OECD, 1999): 73–78.

Iravani, Siavash. "Green synthesis of metal nanoparticles using plants," *Green Chemistry* 13 (2011): 2638–2650.

Jas, Nathalie and Soraya Boudia. *Toxicants, Health and Regulation since 1945* (Cambridge: Cambridge U. Press, 2013).

Jaussaud, Philippe. *Chimie verte: de la plante au médicament* (Paris: SUTIP SA, 1992).

Guinée, Jeroen B., Reinout Heijungs, Gjalt Huppes, Alessandra Zamagni, Paolo Masoni, Roberto Buonamici, Tomas Ekvall, Tomas Rydberg. "Life Cycle Assessment: Past, Present, and Future," *Environmental Science & Technology* 45 (2011): 90–96.

Jessop, Philip. "Searching for green solvents," *Green Chemistry* 13 (2011): 1391.

Jessop, Philip. "Editorial: Evidence of a significant advance in green chemistry," *Green Chemistry* 22 (2020): 13–15.

Keijer, Tom, Vincent Bakker and J. Chris Slootwe.g. "Circular chemistry to enable a circular economy," *Nature Chemistry* 11 (2019): 190–195.

Kettering, C. F. "Studying the knocks," *Scientific American*, 11 October, 1919: 364.

Kharissova, Oxana V., H. V. Rasika Dias, Boris I. Kharisov, Betsabee Olvera Pérez, Victor M. Jiménez Pérez, "The greener synthesis of nanoparticles," *Trends in Biotechnology* 31 (2013): 240–248.

Kirschner, M. "Zauberstoff für eine Sanfte Chemie," *Bild der Wissenschaft* 4, 1993: 14–18.

Klein, Burton H. "The Public Policy Issues Involved in Dealing with Environmental Degradation: A Dynamic Approach," in John O'M. Bockris, ed., *Environmental Chemistry* (New York: Plenum Press, 1978), 749–769.

Krasnodębski, Marcin. "*From Forest Waste to Fuel Tank: The French Quest for Autarkic Fuel Sources after World War I*," *Technology and Culture*, 62 (2021): 105–127.

Krasnodębski, Marcin. "Institut du Pin et la chimie des resines en Aquitaine (1900–1970)" (PhD diss., University of Bordeaux, 2016).

Krasnodębski, Marcin. "Lost green chemistries: history of forgotten environmental trajectories," *Centaurus* (accepted).

Krasnodębski, Marcin. "The Social Construction of Pine Forest Wastes in Southwestern France During the Nineteenth and Twentieth Centuries," *Environment and History* 28 (2022): 155–183.

Kuhn, Thomas S. *The Structure of Scientific Revolutions* (Chicago: University of Chicago Press, 1962).

Kümmerer Klaus, Ann-Kathrin Amsel Amsel, and Dorota Bartkowiak. "Key Characteristics of Sustainable Chemistry. Towards a Common Understanding of Sustainable Chemistry," published in on 13/01/2021 on the ISC3's website (accessed 23/02/2022): https://www.isc3.org/fileadmin/user_upload/Documentations_Report_PDFs/ISC3_Sustainable_Chemistry_key_characteristics_20210113.pdf.

Kümmerer, Klaus and James Clark. "Green and Sustainable Chemistry" in *Sustainability Science. An Introduction*, ed. Harald Heinrichs, Pim Martens, Gerd Michelsen, and Arnim Wiek (Dordrecht: Springer, 2016): 43–60.

Kümmerer, Klaus, James H. Clark, and Vânia G. Zuin. "Rethinking chemistry for a circular economy," *Science* 367 (2020): 369–370.

Kümmerer, Klaus. "Sustainable Chemistry: A Future Guiding Principle," *Angewandte Chemie – International Edition* 56 (2017): 16420–16421.

Laforest, Valérie. "Applying Best Available Techniques in Environmental Management Accounting: From the Definition to an Assessment Method," in

Environmental Management Accounting for Cleaner Production, ed. Stefan Schaltegger, Martin Bennett, Roger L. Burritt and Christine Jasch (Springer, 2008): 29–47.

Lancaster, Mike. *Green chemistry: an introductory text* (Cambridge: Royal Society of Chemistry, 2002).

Lange, Cristoph. *Umweltschutz und Unternehmensplanung. Die betriebliche Anpassung an den Einsatz umweltpolitischer* (Wiesbaden: Gabler KG, 1978).

Lankey, Rebecca L. and Paul Anastas, ed., *Advancing Sustainability through Green Chemistry and Engineering* (Washington D.C.: American Chemical Society, 2002).

Lapkin, Alexei and David J. C. Constable, ed. *Green chemistry metrics. Measuring and Monitoring Sustainable Processes* (London: Wiley-Blackwell, 2008).

Lasker, Grace A. and Edward J. Brush. "Integrating social and environmental justice into the chemistry classroom: a chemist's toolbox," *Green Chemistry Letters and Reviews* 12 (2019): 168–177.

Lasker, Grace A., Karolina E. Mellor, Melissa L. Mullins, Suzanne M. Nesmith, and Nancy J. Simcox. "Social and Environmental Justice in the Chemistry Classroom," *Journal of Chemical Education* 94 (2017): 983–987.

Leitner, Walter, "The subject of 'fracking' in Green Chemistry," *Green Chemistry*, 17 (2015): 2609.

Lenoir, Dieter, Karl-Werner Schramma, Joseph O. Lalah. "Green Chemistry: Some important forerunners and current issues," *Sustainable Chemistry and Pharmacy* 18 (2020): 100313.

Li, Chao-Jun. "Reflection and perspective on green chemistry development for chemical synthesis—Daoist insights," *Green Chemistry* 18 (2016): 1836–1838.

Li, Xiaohong. *Industrial Ecology and Industry Symbiosis for Environmental Sustainability: Definitions, Frameworks and Applications* (London: Palgrave Pivot, 2018).

Linder, Mats. "Ripe for disruption: reimagining the role of green chemistry in a circular economy," *Green Chemistry Letters and Reviews* 10 (2017): 428–435.

Linkwitz, Michael and Ingo Eilks. "Green Chemistry in der Schule," *Chemie in unserer Zeit* 53 (2019): 412–420.

Linthorst, J. A. "An overview: origins and development of green chemistry," *Foundations of Chemistry* 12 (2010): 55–68.

Livage, Jacques and Thibaud Coradin, "Le verre biologique inspire les chimistes," *Pour la science* 371 (2008): 30–34.

Livage, Jacques. "Chimie douce: from shake-and-bake processing to wet chemistry," *New Journal of Chemistry* 25 (2001): 1.

Livage, Jacques. "Vers une chimie écologique. Quand l'air et l'eau remplacent le pétrole," *Le Monde*, 26 October, 1977.

Llored, Jean-Pierre and Stephane Sarrade. "Connecting the philosophy of chemistry, green chemistry, and moral philosophy," *Foundations of Chemistry* 18 (2016): 125–152.

Llored, Jean-Pierre. "Towards a Practical Form of Epistemology: the Case of Green Chemistry," *Studia Philosophica Estonica* 5 (2012): 36–60.

Loste, Natalia, Esther Roldán, and Beatriz Giner. "Is Green Chemistry a feasible tool for the implementation of a circular economy?," *Environmental Science and Pollution Research* 27 (2020): 6215–6227.

Mackenzie, Debora. "Technology: Italian firm first with 'truly biodegradable' plastic," *New Scientist* (25 November 1989).

Mahaffy, Peter G., Edward J. Brush, Julie A. Haack, and Felix M. Ho. "Journal of Chemical Education Call for Paper Special Issue on Reimagining Chemistry Education: Systems Thinking, and Green and Sustainable Chemistry," *Journal of Chemical Education* 95 (2018): 1689–1691.

Maisonobe, Marion, Bastien Bernela, "Exploring the borders of a transregional knowledge network. The case of a French research federation in green chemistry," 2019, working paper hal-02053595 (accessed 22/02/2022): https://hal.archives-ouvertes.fr/hal-02053595/file/MM%26BB.pdf

Malerbe, Arnaud. *La chimie verte: Quelles stratégies pour les industries du sucre et de l'amidon* (Grignon: INRA, 1990).

Malle, Karl-Geert. "Sanfte Chemie halbokkult?," *Nachrichten aus Chemie, Technik und Laboratorium* 42 (1994): 64.

Manahan, Stanley. *Green Chemistry and ten commandments of sustainability* (Columbia: ChemChar Research, 2006).

Manahan, Stanley. *Environmental Chemistry* (Boca Raton: CRC Press, 2017).

Manahan, Stanley. *Industrial Ecology: Environmental Chemistry and Hazardous Waste* (Boca Raton: CRC, 1999).

Marcelino, Leonardo Victor, Adilson Luiz Pinto, and Carlos Alberto Marques, "Intellectual authorities and hubs of Green Chemistry," *TransInformação* 32 (2020).

Marion, P, B. Bernela, A. Piccirilli, B. Estrine, N. Patouillard, J. Guilbotf and F. Jérôme. "Sustainable chemistry: how to produce better and more from less?," *Green Chemistry* 19 (2017): 4973–4989.

Marteel-Parrish, Ann, and Heather Harvey. "Applying the principles of green chemistry in art: design of a cross-disciplinary course about 'art in the

Anthropocene: greener art through greener chemistry,'" *Green Chemistry Letters and Reviews* 12 (2019): 147–160.

Massard-Guilbaud, Geneviève. *Histoire de la pollution industrielle - France, 1789-1914* (Paris: EHESS, 2010).

Matheru, Avtar and Pascale Champagne. "Brown to green and sustainable chemistry," *Current Opinion in Green and Sustainable Chemistry* 2 (2016): iii-iv.

Matlack, Albert. "Teaching green chemistry," *Green Chemistry* 1 (1999): 19–20.

Matlack, Albert. *Introduction to Green Chemistry* (New York: Dekker, 2001).

Matlin, Stephen, Goverdhan Mehta, Henning Hopf, and Alain Krief. "One-world chemistry and systems thinking," *Nature Chemistry* 8 (2016): 393–398.

Matlin, Stephen. Sustainability and chemistry, 2020, published on the IOCD's website (accessed 23/02/2022): http://www.iocd.org/perspectives/606-2020-IOCDperspectve-SustainabilAndChem-13p.pdf

Maxim, Laura, ed. *Chimie Durable: Au-delà des promesses* (Paris: CNRS Editions, 2011).

Mohan, S. Venkata and Ranaprathap Katakojwala. "The circular chemistry conceptual framework: A way forward to sustainability in industry 4.0," *Current Opinion in Green and Sustainable Chemistry* 28 (2021): 100434.

Mohanty, Ajit K., Manjusri Misra & Lawrence T. Drzal. "Sustainable Bio-Composites from renewable resources: Opportunities and challenges in the green materials world," *Journal of Polymers and the Environment* 10 (2002): 19–26.

Moore, John and Elizabeth A. Moore, *Environmental Chemistry* (New York: Academic Press, 1976).

Mudring, Anja-Verena. "Editorial," *Green Chemistry Letters and Reviews* 14 (2021): 1.

Müller, H. "Sanfte Chemie," *Nachrichten aus Chemie, Technik und Laboratorium* 36 (1988): 1011.

Murphy, Mark A. "Early Industrial Roots of Green Chemistry and the history of the BHC Ibuprofen process invention and its Quality connection," *Foundations of Chemistry* 20 (2018): 121–165.

Namaroff, Tamara J., R. J. Garant, M. B. Albert. "Adoption of Green Chemistry: an analysis based on U.S. patents," *Research Policy* 33 (2004): 959–974.

Náray-Szabó, Gábor. "Conservative Evolution, Sustainability, and Culture," *Comparative Literature and Culture*, 16 (2014): http://dx.doi.org/10.7771/1481-4374.2316

Náray-Szabóa, Gábor and László T. Mika. "Conservative Evolution and Industrial Metabolism in Green Chemistry," *Green Chemistry* 20 (2018): 2171–2191.

National Research Council. *Sustainability in the Chemical Industry, Grand Challenges and Research Needs* (Washington D.C.: The National Academies Press, 2006).

Nieddu, Martino, Franck-Dominique Vivien. "La chimie verte: une fausse rupture? Les trajectoires d'une transition écologique," *La Découverte* 2 (2015): 139–153.

Nieddu, Martino, Franck-Dominique Vivien, Estelle Garnier, and Christophe Bliard. "Existe-t-il réellement un nouveau paradigme de la chimie verte?," *Natures Sciences Sociétés* 22 (2014): 103–113.

Nieddu, Martino. "Bernadette Bensaude-Vincent et Jonathan Simon, 2008, Chemistry, The Impure Science, London, Imperial College Press, p. 268 + index.," *Développement durable et territoires* [On line], Lectures (2002–2010), 2008 (accessed 22/02/2022): http://journals.openedition.org/developpementdurable/8216

No author, untitled, *Scientific American*, 3 May, 1919, 459 and 474.

No author. "Editorial," *Washington Post*, 5 May, 1906, 1.

No author. "Executive Summary," *Proceedings of the OECD workshop on sustainable chemistry, Series on Risk Management OECD* (Paris: OECD, 1999): 13–14.

No author. "Ford Predicts Fuel from Vegetation," *New York Times*, 20 September, 1925, 24.

No author. "Launching of a Great Industry: The Making of Cheap Alcohol," *The New York Times*, 25 November, 1906, 3.

No author. "Potential for hemp as green fibre source," *Green Chemistry*, 1 (1999): G159.

No author. "Soya bean glue," *Green Chemistry* 1 (1999): G3–G5.

No author. *Sustainable America: A New Consensus for Prosperity, Opportunity, and a Healthy Environment for the Future* (Washington D.C.: National Science and Technology Council, 1996).

Okkerse, Cédric and Herman van Bekkum. "From Fossil to Green," *Green Chemistry* 2 (1999): 107–114.

Ourisson, Guy. "Chimie polluante, chimie non polluante, chimie dépolluante," in *La chimie, Université de tous les savoirs*, volume 18 (Paris: Odile Jacob, 2002): 219–227.

Pasquon, Italo and Luciano Zanderighi. *La chimica verde: le utilizzazioni dei prodotti vegetali e le biotecnologie* (Milano: Hoepli, 1987).

Pavel Drašar. "Zelená Chemie: Sen nebo Realita (Minimum Impact Chemistry)," *Chemicke Listy* 85 (1991): 1144–1149.

Pena, David S. "The Six Essential Components of Sustainable Socialism: From Building the Productive Forces to Combating Bourgeois Liberalization," *International Critical Thought* 4 (2014): 267–288.

Peplow, Mark. "Warning Shot for Green Chemists: Some Solvents with an Environmentally Friendly Reputation May Kill Fish," *Nature* (3 November 2005), https://doi.org/10.1038/news051031-8.

Pétain, Philippe. "Carburant national et véhicules à gazogène," Revue hebdomadaire 17 (1936): 391–402.

Pierre Gazellot. "Conversion of biomass to selected chemical products," *Chemical Society Reviews* 41 (2012): 1538–1558.

Poliakoff, Martyn, J. Micheal Fitzpatrick, Trevor Farren and Paul Anastas. "Green chemistry: Science and politics of change," *Science* 297, No. 5582 (2002): 807–810.

Poliakoff, Martyn, Walter Leitner and Emilia S. Strengab. "The Twelve Principles of CO2 CHEMISTRY," *Faraday Discussions* 183 (2015): 9–17.

Poliakoff, Martyn. "Kletz T. A new father figure?," *TCE: The Chemical Engineer*, No. 866 (2013): 42–37.

Reniers, Genserik and Carlos A. Brebbia. *Sustainable Chemistry* (Antwerp: WIT Press, 2011).

Ribeiro M. Gabriela T.C. and Adélio A.S.C. Machado. "Greenness of chemical reactions – limitations of mass metrics," *Green Chemistry Letters and Reviews* 6 (2013): 1–18.

Ribeiro, M. Gabriela T. C., Santiago F. Yunes, and Adélio A. S. C. Machado, "Assessing the Greenness of Chemical Reactions in the Laboratory Using Updated Holistic Graphic Metrics Based on the Globally Harmonized System of Classification and Labeling of Chemicals," *Journal of Chemical Education* 91 (2014): 1901–1908.

Ribeiro, M. Gabriela T.C., Dominique A. Costa & Adélio A.S.C. Machado. "'Green Star:' a holistic Green Chemistry metric for evaluation of teaching laboratory experiments," *Green Chemistry Letters and Reviews* 3 (2010): 149–159.

Roberts, Jody. "Creating Green Chemistry: Discursive Strategies of a Scientific Movement" (PhD diss., Faculty of Virginia Polytechnic Institute and State University, 2006).

Rockström, Johan, Will Steffen, Kevin Noone, Asa Persson, F Stuart Chapin 3rd, Eric F Lambin, Timothy M Lenton, Marten Scheffer, Carl Folke, Hans Joachim Schellnhuber, Björn Nykvist, Cynthia A. de Wit, Terry Hughes, Sander van der Leeuw, Henning Rodhe, Sverker Sörlin, Peter K. Snyder, Robert Costanza, Uno Svedin, Malin Falkenmark, Louise Karlberg, Robert W. Corell, Victoria J. Fabry, James Hansen, Brian Walker, Diana Liverman, Katherine

Richardson, Paul Crutzen, Jonathan A Foley. "A safe operating space for humanity," *Nature* 461 (2009): 472–475.

Roger, Jacques. *Pour une histoire des sciences à part entière, Texte établi par Claude Blanckaert* (Paris: Albin Michel, 1995).

Rogers, Robin and Kenneth Seddon. *Ionic Liquids, Industrial Applications for Green Chemistry* (Oxford: Oxford University Press, 2002).

Roon, André van, Harrie A.J. Govers, R. John and Hans van Weenen. "Sustainable chemistry: an analysis of the concept and its integration in education," *Journal of Sustainability in Higher Education*, 2 (2001): 161–180.

Roon, André van. "Designing sustainable chemicals: predictive tools for the environmental fate of monoterpene pesticide" (Ph.D. diss. University of Amsterdam, 2006).

Rotman, David. "Chemists map greener synthesis pathways," *Chemical Week* 153 (1993): 56–57.

Sanchez, Clement, Laurence Rozes, François Ribot, C. Laberty-Robert, D. Grosso, C. Sassoye, C. Boissiere, L. Nicole. "'Chimie douce:' A land of opportunities for the designed construction of functional inorganic and hybrid organic-inorganic nanomaterials," *Comptes Rendus Chimie* 13 (2010): 3–39.

Sarrade, Stéphane. *La chimie d'une planète durable* (Paris: Le Pommier, 2011).

Sarrade, Stéphane. *Ressources de la Chimie Verte* (Paris: EDP, 2008).

Sato, K., M. Aoki, and R. Noyori. "A "Green" route to adipic acid: direct oxidation of cyclohexenes with 30 percent hydrogen peroxide," *Science* 281 (1998): 1646–1647.

Sawyer Donald T. and Arthur E. Martell, ed. *Industrial Environmental Chemistry: Waste Minimization in Industrial Processes and Remediation of Hazardous Waste* (New York: Springer, 1992).

Scharfe, G., B. Seweko, "Produktionsintegrierter Umweltschutz – Das Beispiel Bayer," *Chemische Industrie, Sonderausgabe Nordrhein-Westfalen* (1991): 17–20.

Servos, John W. *Physical Chemistry from Ostwald to Pauling* (Princeton: Princeton University Press, 1990).

Sheldon Roger A. and John M. Woodley. "Role of Biocatalysis in Sustainable Chemistry," *Chemical Reviews* 118 (2018): 801–838.

Sheldon Roger A. and Michael Norton. "Green chemistry and the plastic pollution challenge: towards a circular economy," *Green Chemistry* 22 (2020): 6310–6322.

Sheldon, Roger A. "Green solvents for sustainable organic synthesis: State of the art," *Green Chemistry* 7 (2005): 267–278.

Sheldon, Roger A. "The E Factor: fifteen years on," *Green Chemistry* 9 (2007): 1273-1283.

Sheldon, Roger A. "Editorial," *Green Chemistry* 2 (2000): G1-G2.

Sheldon, Roger A. "Green and Sustainable Manufacture of Chemicals from Biomass: State of the Art," *Green Chemistry* 16 (2014): 950-963.

Sheldon, Roger A. "Green chemistry and resource efficiency: towards a green economy," *Green Chemistry* 18 (2016): 3180-3183.

Sheldon, Roger A. "The E Factor: fifteen years on," *Green Chemistry* 9 (2007): 1273-1283.

Simmons, Milagros S. "The Role of Catalysts in Environmentally Benign Synthesis of Chemicals," in *Green Chemistry. Designing Chemistry for the Environment*, ed. Paul Anastas and Tracy C. Williamson (Washington D.C.: American Chemical Society, 1996), 116-130.

Sjöström, Jesper. "Green chemistry in perspective—models for GC activities and GC policy and knowledge areas," *Green Chemistry* 8 (2006): 130-137.

Smith, David J. "A Fresh Look at Sustainable Chemistry," *ChemSusChem* 14 (2021): 5-9.

Snow, Bradley D. *Living with Lead: An Environmental History of Idaho's Coeur D'Alenes, 1885-2011* (Pittsburgh: University of Pittsburgh Press, 2017).

Steinhäuser, Klaus Günter, Steffi Richtera and Jutta Penninga, "Sustainable Chemistry in Dessau – a workshop report," *Green Chemistry* 6 (2004): G41-G43.

Swatloski, Richard P., John D. Holbrey, and Robin D. Rogers. "Ionic liquids are not always green: hydrolysis of 1-butyl-3-methylimidazolium hexafluorophosphate," *Green Chemistry* 5 (2003): 361-363.

Tang, Samantha L. Y., Richard A. Bourne, Richard L. Smith, and Martyn Poliakoff. "The 24 Principles of Green Engineering and Green Chemistry: "IMPROVEMENTS PRODUCTIVELY,"" *Green Chemistry* 10 (2008): 268-269.

Tang, Samantha L. Y., Richard L. Smith, and Martyn Poliakoff. "Principles of green chemistry: PRODUCTIVELY," *Green Chemistry* 7 (2005): 761-762.

Teissier, Pierre. "Une histoire de la chimie du solide" (PhD diss. University of Nantes, 2008).

Teissier, Pierre. *Une histoire de la chimie du solide* (Paris: Herrman, 2014).

Thornton, Joe. *Pandora's Poison, Chlorine, Health, and a New Environmental Strategy* (Cambridge, MA: MIT Press, 2000).

Tonkiss, Fran. "Discourse Analysis," in *Researching Society And Culture*, ed. Clive Seale (London: Sage, 2004), 245-260.

Travers, A. "La Crise des Carburants les rémèdes proposés," *Bulletin de l'Institut du Pin* 17 (1925): 241-243.

Trost, Barry. "The Atom Economy – A Search for Synthetic Efficiency," *Science* 254 (1991): 1471–1477.

Tundo, Pietro, Paul Anastas, David StC. Black Joseph Breen, Terrence Collins, Sofia Memoli, Junshi Miyamoto, Martyn Polyakoff, and William Tumas. "Synthetic pathways and processes in green chemistry. Introductory overview," *Pure and Applied Chemistry* 72 (2000): 1207–1228.

Tundo, Pietro. "Green Chemistry on the Rise," *Chemistry International* 29 (2007): 5–7.

Voillequin, Baptiste. *La catalyse en France (1944-2004): Dynamiques disciplinaires et régimes de production de savoir* (Paris: Editions universitaires européennes, 2010).

Waked, Alexander E., Karl Z. Demmans, Rachel F. Hems, Laura M. Reyes, Ian Mallov, Erika Daley, Laura B. Hoch, Melanie L. Mastronardi, Brian J. De La Franier, Nadine Borduas, and Andrew P. Dicks. "The Green Chemistry Initiative's Contributions to Education at the University of Toronto and Beyond," *Green Chemistry Letters and Reviews* 12 (2019): 187–195.

Wang, Yong, Xinchen Wang, and Markus Antonietti. "Polymeric graphitic carbon nitride as a heterogeneous organocatalyst: From photochemistry to multipurpose catalysis to sustainable chemistry," *Angewandte Chemie – International Edition* 51 (2012): 68–89.

Warner, John. "The natural evolution of green chemistry," *Green Chemistry Letters and Reviews* 1 (2007): 1–2.

Wasserscheid, Peter, Annegret Stark, Paul Anastas, ed. *Green Solvents: Ionic Liquids*, (Wiley-VCH, 2013).

Weise, Eherhard, Henning Friege, Karl Otto Henseling, Ian C. Meerkarnp van Embden, Bud Homburg. "Wie die Chemie "grün" wurde," *Nachrichten aus Chemie, Technik und Laboratorium* 47 (1999): 914–918.

Welton, Tom. "Ionic liquids in Green Chemistry," *Green Chemistry* 13 (2011): 225.

White, Allen and Diana Zinkl. *Green Metrics: A Status Report on Standardized Corporate Environmental Reporting* (Boston: Tellus Institute, 1997).

Wilkes, John S. "A short history of ionic liquids—from molten salts to neoteric solvents," *Green Chemistry* 4 (2002): 73–80.

Winterton, Neil. "Chlorine: the only green element – towards a wider acceptance of its role in natural cycles," *Green Chemistry* 2 (2000): 173–225.

Winterton, Neil. "Twelve more green chemistry principles," *Green Chemistry* 3 (2001): G73–G75.

Woodhouse Edward J. and Steve Breyman. "Green Chemistry as Social Movement?," *Science, Technology, & Human Values* 30 (2005): 199–222.

Zhang, Qinghua, Karine De Oliveira Vigier, Sébastien Royer and François Jérôme. "Deep eutectic solvents: syntheses, properties and applications," *Chemical Society Reviews* 41 (2012): 7108–7146.

Zhang, Xi-Feng, Zhi-Guo Liu, Wei Shen, Sangiliyandi Gurunathan, "Silver Nanoparticles: Synthesis, Characterization, Properties, Applications, and Therapeutic Approaches," *International Journal of Molecular Sciences* 17 (2016): 1534.

Zhu, Hongli, Wei Luo, Peter N. Ciesielski, Zhiqiang Fang, J Y Zhu, Gunnar Henriksson, Michael E Himmel, Liangbing Hu. "Wood-Derived Materials for Green Electronics, Biological Devices, and Energy Applications," *Chemical Reviews* 116 (2016): 9305–9374.

Zuin, Vânia G., Ingo Eilks, Myriam Elschami and Klaus Kümmerer. "Education in green chemistry and in sustainable chemistry: perspectives towards sustainability," *Green Chemistry* 23 (2021): 1594–1608.

Zuin, Vânia G., Mateus L. Segatto, Dorai P. Zandonai, Guilherme M. Grosseli, Aylon Stahl, Karine Zanotti, Rosivania S. Andrade. "Integrating Green and Sustainable Chemistry into Undergraduate Teaching Laboratories: Closing and Assessing the Loop on the Basis of a Citrus Biorefinery Approach for the Biocircular Economy in Brazil," *Journal of Chemical Education* 96 (2019): 2975–2983.

Websites:

12 Principles of Green Chemistry on the official website of the American Chemical Society (accessed 23/05/2022): https://www.acs.org/content/acs/en/greenchemistry/principles/12-principles-of-green-chemistry.html

Article in the national Italian encyclopaedia Treccani (accessed 18/02/2022): https://www.treccani.it/enciclopedia/italo-pasquon.

Booklet on various types of green chemistry principles (accessed 19/02/2022): https://www.acs.org/content/acs/en/greenchemistry/principles/design-principles-booklet.html

Ellen McArthur Foundation's website, circular economy section (accessed 23/02/2022): https://ellenmacarthurfoundation.org/towards-the-circular-economy-vol-1-an-economic-and-business-rationale-for-an

European Parliament's website, circular economy definition (accessed 23/02/2022): https://www.europarl.europa.eu/news/en/headlines/economy/20151201STO05603/circular-economy-definition-importance-and-benefits

Hague Ethical Guidelines (accessed 23/02/2022): https://www.opcw.org/hague-ethical-guidelines

IOCD's website, mission and vision section (accessed 23/02/2022): http://www.iocd.org/About/visionMission.shtml

IUPAC's website, project "Learning Objectives and Strategies for Infusing Systems Thinking into (Post)-Secondary General Chemistry Education" (accessed 23/02/2022): https://iupac.org/projects/project-details/?project_nr=2017-010-1-050

Michael Fumento's "Rachel's Folly" from 1996, (accessed 20/02/2022): http://lobby.la.psu.edu/015_Disinfectant_Byproducts/Organizational_Statements/C3/C3_The_End_of_Chlorine.htm

Official website of the company AURO (accessed 22/02/2022): https://www.auro.de/de/ueber-AURO/sanfte-chemie/fachbeitraege/

Report "Mit Nachwachsenden Rohstofen auf dem Weg zur Nachhaltigkeit," *Forschungsforum* 3 (1997) (accessed 22/02/2022): https://nachhaltigwirtschaften.at/resources/nw_pdf/fofo/fofo3_97_de.pdf

Roger Sheldon's personal website (accessed 19/02/2022): https://www.sheldon.nl/roger/efactor.html.

Scopus website support hub (accessed 21/02/2022): https://service.elsevier.com/app/answers/detail/a_id/21730/supporthub/scopus/

The first instance of Wikipedia article on environmental chemistry mentioning green chemistry in a slightly modified form comes from 23 January 2005 (accessed 18/02/2022): https://en.wikipedia.org/w/index.php?title=Environmental_chemistry&diff=11018353&oldid=9584656.

The Green Chemistry award's official website, 1996 winners (accessed 19/02/2022): https://www.epa.gov/greenchemistry/green-chemistry-challenge-winners#1996.

The modification of the Green Chemistry article in Wikipedia from 23 November 2007 (accessed 22/02/2022): https://en.wikipedia.org/w/index.php?title=Green_chemistry&diff=173285727&oldid=173250886

The official website of the ISC3, history section (accessed 22/02/2022): https://www.isc3.org/en/about-isc3/history.html.

The official website of the ISGC symposia (accessed 22/02/2022): https://www.isgc-symposium.com/

The teaching curriculum of the Milan Polytechnic in 1996–1996 (accessed 18/02/2022): https://www.ingindinf.polimi.it/fileadmin/user_upload/scuola/programmi_insegnamenti_veccio_ordinamento/1995-96.pdf, p. 57.

U.S. Patent Statistics Chart Calendar Years 1963–2020 (accessed 21/02/2022), https://www.uspto.gov/web/offices/ac/ido/oeip/taf/us_stat.htm

Website of the 1996 Gordon Research Conference in green chemistry (accessed 26/02/2022): https://www.grc.org/environmentally-benign-organic-synthesis-conference/1996/

Website of the German environmental think tank Katalyse Institute, accessed 22/02/2022, http://umweltlexikon.katalyse.de/?p=1115

Wikipedia article on green engineering (accessed 19/02/2022): https://en.wikipedia.org/wiki/Green_engineering

World Commission on Environment and Development, "Our Common Future" (accessed 22/02/2022): https://sustainabledevelopment.un.org/content/documents/5987our-common-future.pdf

Index of Names

A
Abraham, M. 103
Amato, I. 63–67
Anastas, P. 29, 31–37, 40–42, 44, 45, 49, 51, 52, 54–63, 65–67, 69–75, 77, 79, 81, 82, 87–96, 98, 100–103, 110–112, 115, 118, 120, 122, 123, 125, 129, 134, 140, 142, 145–147, 149, 150, 152, 156, 157, 166, 168, 169, 176, 178, 179, 181–185, 205, 209, 210, 212, 213, 215–217, 223, 224, 227, 234, 236, 252, 254, 258, 261

B
Bashkin, J. 119, 120, 124
Berzelius, J. 40
Bloch, E. 198
Bockris, J.O'M. 262
Böschen, S. 221, 226, 227
Bozell, J. 156, 185
Breen, J. 45, 72, 73, 79, 142, 209, 212, 213, 215
Breyman, S. 26, 27, 31, 39, 63, 69, 71, 72, 98, 265
Brindle, I. 84, 216
Brush, E. 259, 260, 265

C
Carra, J. 215
Carson, R. 33, 37, 40, 43, 121–123, 168, 206
Cavanaugh, M. 65–67, 73
Chardin, T. 245, 246
Christ, C. 201, 204, 205, 221, 222, 239, 272

Clark, J. 36, 38, 39, 51, 79–83, 123, 125, 132, 140, 142, 145–147, 149, 157, 162, 165, 177, 181, 183–187, 189–191, 199, 232, 256, 258
Collins, T. 85–88, 121, 222
Colonna, P. 177, 178, 181
Constable, D. 75, 101, 130–133, 166, 169, 170

D
Deetlefs, M. 126, 127
DeVito, S. 68–70, 90, 91
Drašar, P. 53, 78
Ducamp, C. 179

F
Finlay, M. 42, 175
Fischer, H. 197–200
Ford, H. 174, 175
Friege, H. 218, 233

G
Garnier, E. 26, 29, 84, 85, 92, 93, 98, 176–178, 180, 182, 184–186, 189, 195, 220
Garrett, R. 68–73, 77, 88–93, 129
Glaze, W. 64, 100
Gleich, A. 197–200, 227
Goethe, J.W. 198
Graedel, T. 211–214, 260, 261

H
Hancock, K. 63–68, 70–73, 77, 87, 88, 129
Holbrey, J.D. 126, 149
Hutzinger, O. 217, 218, 228

J
Jessop, P. 128, 133, 134

K
Kettering, C. 175
Kitazume, T. 142
Kletz, T. 40–42
Korte, F. 221, 227
Kuhn, T. 25, 246
Kümmerer, K. 135, 161, 204, 228, 231, 232, 234, 237, 239, 256

L
Lancaster, M. 130, 147
Lasker, G. 265, 267
Linthorst, J. A. 26, 27, 31, 56, 57, 79, 80, 98, 140
Livage, J. 195–197
Llored, J. 26, 178, 179
Loupy, A. 149
Lovelock, J. 245

M
Macquarrie, D. 39, 142, 146, 147, 184
Manahan, S. 109, 111, 209, 222
Marcuse, H. 198
Martell, A. 78
Matlack, A. 87, 88, 95, 111
Matlin, S. 246, 248, 249
Maxim, L. 179
Molin, M. 64
Moore, E. 261
Moore, J. 261
Mudring, A.-V. 117, 118
Murphy, M. 41, 46, 168, 264, 265

N
Nguyen, N. 103
Nieddu, M. 26, 29, 92, 98, 176–178, 180, 185, 186, 195
Nobel, A. 40–42

P
Pasquon, I. 50–52, 176, 191
Pétain, P. 174, 175
Poliakoff, M. 32, 36, 41, 49, 105, 107, 108, 112, 149, 189, 223

R
Reniers, G. 229
Rico-Lattes, I. 178
Roberts, J. 26–29, 31, 37, 38, 44, 45, 49, 50, 54, 56, 61–66, 92, 97, 98, 111, 112, 122, 123, 126, 177
Roger, J. 25
Rogers, R. 32, 125, 126, 147, 149
Roosevelt, T. 174
Rouxel, J. 195, 196

S
Sarrade, S. 26, 178–180
Sawyer, D. 78
Seddon, K. 32, 125–127, 145, 147
Sheldon, R. 14, 53, 74–78, 106, 129–131, 150, 164, 189, 190, 222, 251, 258
Sjöström, J. 111, 186, 187
Steiner, R. 198

T
Trost, B. 74, 75, 77, 78, 92, 106, 129

W
Walden, P. 40
Warner, J. 31–36, 41, 44, 45, 51, 54, 55, 61, 65, 74, 77, 79, 81, 88, 93, 96, 100, 101, 111, 125, 146, 147, 152, 166, 168, 176, 182–184, 210, 213, 215, 224, 234, 236, 252, 258, 261
Williamson, T. 52, 56, 59–61, 73, 77, 89, 92, 134, 142, 147
Winterton, N. 43, 99, 100, 101, 103–105, 118, 121, 122, 186

Woodhouse, E. 26, 27, 31, 38, 39, 42, 51, 63, 69, 71, 72, 97, 98, 184, 199, 264, 265

Z

Zanderighi, L. 50–52, 176, 191
Zimmerman, J. 32, 101–103
Zuin, V. 108, 114, 204, 239, 256

Polish Contemporary Philosophy and Philosophical Humanities

Edited by Jan Hartman

Vol. 1 Roman Murawski: Logos and Máthēma. Studies in the Philosophy of Mathematics and History of Logic. 2011.

Vol. 2 Cezary Józef Olbromski: The Notion of *lebendige Gegenwart* as Compliance with the Temporality of the "Now". The Late Husserl's Phenomenology of Time. 2011.

Vol. 3 Jan Woleński: Essays on Logic and its Applications in Philosophy. 2011.

Vol. 4 Władysław Stróżewski: Existence, Sense and Values. Essays in Metaphysics and Phenomenology. Edited by Sebastian Kołodziejczyk. 2013.

Vol. 5 Jan Hartman: Knowledge, Being and the Human. Some of the Major Issues in Philosophy. Translated by Ben Koschalka. 2013.

Vol. 6 Roman Ingarden: Controversy over the Existence of the World. Volume I. Translated and annotated by Arthur Szylewicz. 2013.

Vol. 7 Jan Hartman: Philosophical Heuristics. Translated by Ben Koschalka. 2015.

Vol. 8 Roman Ingarden: Controversy over the Existence of the World. Volume II. Translated and annotated by Arthur Szylewicz. 2016.

Vol. 9 Tomasz Kubalica: Unmöglichkeit der Erkenntnistheorie. Leonard Nelsons Kritik an der Erkenntnistheorie unter besonderer Berücksichtigung des Neukantianismus. 2017.

Vol. 10 Renata Ziemińska: The History of Skepticism. In Search of Consistency. 2017.

Vol. 11 Jan Woleński: Logic and Its Philosophy. 2018.

Vol. 12 Wielslaw Gumula: On Property and Ownership Relations. 2018.

Vol. 13 Andrzej Zaporowski: Action, Belief, and Community. 2018.

Vol. 14 Andrzej Bator / Zbigniew Pulka (eds.): A Post-Analytical Approach to Philosophy and Theory of Law. 2019.

Vol. 15 Krzysztof Śleziński: Towards Scientific Metaphysics. Volume 1: In the Circle of the Scientific Metaphysics of Zygmunt Zawirski. Development and Comments on Zawirski's Concepts and their Philosophical Context. 2019.

Vol. 16 Krzysztof Śleziński: Towards Scientific Metaphysics. Volume 2: Benedykt Bornstein's Geometrical Logic and Modern Philosophy. A Critical Study. 2019.

Vol. 17 Jan Felicjan Terelak: Eustress and Distress: Reactivation. 2019.

Vol. 18 Roman Murawski: Lógos and Máthēma 2. Studies in the Philosophy of Logic and Mathematics. 2020.

Vol 19 Andrzej J. Noras: Geschichte des Neukantianismus. Übersetzt von Tomasz Kubalica. 2020.

Vol. 20 Tomasz Jarmużek: Tableau Methods for Propositional Logic and Term Logic. Translated by Sławomir Jaskólski. 2020.

Vol. 21 Marta Kudelska: Why Is There I Rather Than It? Ontology of the Subject in the Upaniṣads. 2021.

Studies in Philosophy, History of Ideas and Modern Societies

Edited by Jan Hartman

Vol. 22 Georg Schmid: The Treachery of the Elites. On Political Discontent. 2021.

Vol. 23 Hanna Urbańska: The Philosophical System of *Śiva Śatakam* and Other *Śaiva* Poems b Nārāyaṇa Guru. In Relation to *Tirumandiram* by Tirumūlar. 2022.

Vol. 24 Marcin Krasnodębski: Green Chemistry. A Brief Historical Critique. 2022.

www.peterlang.com

www.ingramcontent.com/pod-product-compliance
Ingram Content Group UK Ltd.
Pitfield, Milton Keynes, MK11 3LW, UK
UKHW041902230426
12049UKWH00002B/17